"森林萌主"自然教育系列丛书

"森林萌主"
STEP INTO GIANT PANDA HABITAT
身边的那些事

大熊猫国家公园都江堰管护总站　编著
DUJIANGYAN STATION OF GIANT PANDA NATIONAL PARK

四川科学技术出版社

编辑委员会

主　　任：周宏亮
副 主 任：蒋礼立　罗　申
编　　委：胡　力　尚　涛　徐　利　泽基拉姆
　　　　　　朱大海　刘　波　苟安然　孙　伟
　　　　　　刘雅梦　何海林　马　睿　修云芳
　　　　　　徐素慧　杨　丽　严思思　郑薇薇　王　民

编写组

主　　编：周宏亮
执行主编：费立松
策　　划：蒋礼立　罗　申
撰　　稿：费立松　严思思　杨　丽　郑薇薇
翻　　译：高　堤　赵益丹　林　芮　杨廷婷
英文审校：黄　炎　高　堤
摄　　影：费立松　尚　涛　朱大海　王建和　费宇翔
　　　　　　何　鑫　严思思　文　龙　程蓉伟
图片提供：图虫创意

Editorial Board

Director: Zhou Hongliang

Deputy Directors: Jiang Lili, Luo Shen

Editors: Hu Li, Shang Tao, Xu Li, Ze Ji Lamu, Zhu Dahai, Liu Bo, Gou Anran, Sun Wei, Liu Yameng, He Hailin, Ma Rui, Xiu Yunfang, Xu Suhui, Yang Li, Yan Sisi, Zheng Weiwei, Wang Min

Writing Group

Chief Editor: Zhou Hongliang

Executive Editor: Fei Lisong

Consuitants: Jiang Lili, Luo Shen

Contributors: Fei Lisong, Yan Sisi, Yang Li, Zheng Weiwei

Translate: Gao Di, Zhao Yidan, Lin Rui, Yang Tingting

English Proofreaders: huang Yan, Gao Di

Photographers: Fei Lisong, Shang Tao, Zhu Dahai, Wang Jianhe, Fei Yuxiang, He Xin, Yan Sisi, Wen Long, Cheng Rongwei

序

近几年，自然教育成为越来越热门和越来越重要的社会话题。之所以如此，是因为人类社会正面临着越来越严重的环境问题，世界各国对于利用自然资源的态度变得越来越谨慎，节能减排、抑制全球变暖、保护生物多样性越来越成为国际社会的主流态度。在我国，"绿水青山就是金山银山"已成为党、国家和社会各界的普遍共识。

与此同时，就像本书作者这样，越来越多的有识之士已清醒地认识到，保护环境要从全民教育做起。只有做好自然教育工作，让环境保护的观念深入人心，才能真正保护好环境，才能真正做到可持续发展。尤为重要的是，青少年是国家和民族的未来，只有将热爱生命、感恩自然、善待环境的意识深植于他们的内心，外化在他们的行动，才能从根本上将环境保护转化为全社会的共同意志。本书在这方面做出了非常有益的尝试，它对自然教育的普及和环境保护都将起到积极的推动作用。

　　大熊猫是最为知名的旗舰物种。旗舰物种的意义不仅仅局限于它在自然保护方面的号召力和吸引力，更在于保护好旗舰物种，就意味着保护好相当大面积的自然环境。从这个意义上讲，讲好《"森林萌主"身边的那些事》，就是细致地体验熊猫生存的自然环境。只有了解它，才会热爱它；只有热爱它，才能保护它。这是一个非常棒的自然教育选题和视角。本书的小读者们，期待你们能够通过阅读，了解生态平衡的意义，了解环境保护的意义，渐渐从自我做起，更要带动身边的每个人，让大家都养成节约水电、不浪费各种资源、爱护野生动植物、保护环境的好意识和好习惯。要知道，爱护绿水青山就是爱护你自己。

中国工程院院士　东北林业大学教授　马建章

Prologue

In recent years, nature education has become an increasingly hot and important topic in society. It is because human society is facing gradually severe environmental problems. Countries around the globe are getting far more cautious towards utilizing natural resources. Conserving energy and reducing emission, curbing global warming and protecting biodiversity have become the mainstream attitude of the international community. In China, "green mountains and clear waters are as valuable as mountains of gold and silver" has become the common belief of the Communist Party of China, and all walks of life in China.

Meanwhile, more people are aware that in order to protect the environment, we must start with universal education. Only education can make people realize how important it is to protect the environment and finally make sustainable development become possible. What's more, teenagers are the future of a nation. Their awareness of loving life, appreciating nature, and taking care of the environment will convert into public common sense. This book has conducted a rewarding experience and shall perform immeasurable promoting effect towards nature education and environmental protection.

The giant panda is the most well-known flagship species. The significance of protection flagship species is not only about its appeal and attraction for nature protection, but also protecting extensive mother nature. In this sense, *Step into Giant Panda Habitat* described the living environment of the panda in detail. Only when understand pandas can people love them and proceed to protect them. This is a brilliant topic and perspective of nature education. To the young readers of this book, I hope you understand the meaning of ecological balance and environmental protection through reading. Starting from yourself and affect people around you to save water and electricity, stop wasting resources, protect wild animals, plants as well as the environment. Please keep in mind, protecting the environment is protecting the human race.

MA Jianzhang
Academician of Chinese Academy of Engineering (CAE)
Professor of Northeast Forestry University

前言

婴儿降生后，第一眼看见的光线，几乎没有谁记得或在意过，总觉得太阳升起和夜幕降临都是那么自然而然。长大以后，日出日落、昼夜交替、风风雨雨人人都习以为常。人类生活在天地之间，蒙受着大自然的恩宠，因此，我们应该去认真了解孕育所有生命的大自然。

人们都珍惜生命，把自己的人生看得很重要。因此，大家热爱生命、热爱生活就应该从热爱大自然开始，希望每个人都要关注自己周围的环境，注意生存环境的变化。在这个科技和经济高度发达的时代，人与人之间、人与自然环境之间的关系越来越密切，彼此相互影响，相互依存，密不可分。

人类走出茫茫森林，其实并没有多长时间。人类的祖先几百万年都一直是在森林里生存、繁衍、进化，历经久远才有了现代人，也才有了现代的文明社会。在这个急速发展的社会，我们每个人都应该回过头去看看，我们人类从哪里来？是否有必要关注环境和爱护森林？森林里的生命是多样的，生存形式是丰富多彩的。这些，可能你都知道，也很了解，但可能你没有在意。从现在开始，希望你多关心、多留意离你并不遥远的森林，特别是生活在高楼林立的都市里的小朋友，我们盛情邀请你们走进大熊猫国家公园，去体会生命相互依存的感动。

本书主要通过介绍大熊猫等代表性物种和川西北地区的相关生境，帮助青少年和有兴趣的成年读者了解这一地域的动植物状况。

人类的事情人类管，和谐、持续、美满的生活需要靠自己去创造与实现。不要一发生传染病就甩锅给野生动物，人们要冷静地思考一下，人类过去与今天都做了什么？人类又是怎么管控和约束自己的？认真想一想，你就会明白的。

Introduction

After a baby was born, almost no one remembered or cared about the light that he/she saw at first sight. The sun always rises and the night always falls. When a person grows up, he/she become accustomed to sunrise and sunset, day and night. As human beings live on the earth and are blessed by nature, we are supposed to understand the nature which gives us lives.

People cherish life and value it a lot. Therefore, love for life can start from love for nature. Hopefully, everyone can pay attention to their surroundings and changes to the environment. In this highly developed time, there is a closer relationship between people, and between people and nature. One interacts with the other.

It didn't take long for humans to walk out of the forest. Human ancestors have lived, multiplied, and evolved in the forest for

millions of years before modern people and modern civilized society exist. In this rapid changing society, we should look back and ask: where are we from? Whether it is necessary to care about the environment? Life in the forest is diverse. You may know it well, but nobody cares. From now on, I wish you to care more about the forest that is not far from you. For children who are living in the high-rise city, you are invited to enter the Giant Panda National Park, to experience the interdependence of life.

This book mainly introduces typical species such as giant pandas and their related habitats in Northwest Sichuan to help young people and interested adults understand the flora and fauna of this region.

Humans take care of human affairs, and a harmonious, continuous and happy life is created and achieved by ourselves. Don't blame wild animals for infectious diseases. People should calmly think about what humans have done in the past and today. How do humans control and restrain themselves? Think about it and you will know.

目录

001　第一章　绿色世界
Chapter One Green world

002　森林的生态效益
　　　Ecological Benefit of the Forest

010　没有花朵和果实的植物
　　　Plants without Flowers and Fruits

018　漫山遍野的艳丽花朵
　　　Gorgeous Flowers

030　植物们的生存关系
　　　Coexistence relationship between plants

040　异样的植物生存方式
　　　A different Way of Life for Plants

047 第二章　节肢动物空间
Chapter Two Arthropod Zone

073 第三章　从水到陆
Chapter Three From Water to Land

083 第四章　竹林隐士
Chapter Four Bamboo Forest Hermit

- 084　竹林隐士的生活
  The life of the hermit in the bamboo forest

- 100　大熊猫的艰难经历
  Difficult experience of giant pandas

115 第五章　珍禽异兽
Chapter Five Exotic Animals

- 116　莺飞凤舞的森林空间
  Auspicious (Harmonious) Forest

- 136　兽类的大舞台
  The Big Stage of Beasts

169 第六章　生命共同体
Chapter Six Community of Life

第一章 Green world 绿色世界
Chapter One

森林的生态效益

Ecological Benefit of the Forest

请你静静地走进森林，不预设概念性的想象，不携带污染环境的物品。沿着森林自然的小路，怀着对生命体的同理心，用心探秘生物的多样性，访问和了解生长在这里的每一种植物吧。

Please walk into the forest quietly, without pre-conceptual imagination, and do not carry objects that pollute the environment. Follow the natural path of the forest, with empathy for living organisms, deliberately explore biodiversity, visit and understand every plant that grows here.

丰富的生境和多样的植物群落

　　清晨，阳光的清辉穿透绿色的原始森林，当你步入这片秀美世界的时候，微风携带着花香的气息扑面而来。踏着松软绵厚的落叶、枯草，沿着山谷溪涧前行，薄薄的雾气慢慢上升再弥漫开来，恍若柔丝飘浮在林中，此刻的你，就如同闯入了一个亦真亦幻的童话世界。

　　青藏高原东部边缘的森林地带，海拔垂直变化大，动植物种类繁多。在这里，高寒的流石滩、高山草甸、原始的针叶林、针阔混交林、高山灌丛交错而生。这片绿色的森林，不仅为生物多样性提供了可贵的基础保障，而且还为人类贡献了优良的生存环境。

Abundant habitat and diverse plant communities

In the early morning, the light of the sun penetrates the virgin forest. When you step into this beautiful green world, the breeze brings the scent of flowers to your face. Stepping on the soft and thick deciduous grass, and walking along the valley stream, the thin mist rises and diffuses, like silk floating in the forest. You are like breaking into a dreamy fairy tale world.

Forests on the eastern edge of the Qinghai-Tibet Plateau have large vertical changes and a wide variety of animals and plants. They are interlaced with alpine rock beaches, alpine meadows, primitive coniferous forests, coniferous and broad-leaved mixed forests, and alpine shrubs. This green forest not only provides a basic guarantee for biodiversity, but also contributes an excellent living environment to mankind

大自然的碳工厂和制氧车间

　　森林是绿的世界，只要温暖的阳光洒向大地，光合作用就会源源不断地进行；森林是一个生物多样性富集的胜地，构建了自己完整的生态体系，始终弥漫着浓浓的神秘感；森林是生物圈内最活跃的一个生命大舞台，自养和异养生物不断演绎着一个个精彩的故事……

　　森林植物通过光合作用产生大量的有机质，同时不断释放出氧气，这些都是人类赖以生存所必要的物质。我们只有走进森林，才能了解生活在这里的动植物们，才能亲身感受森林这个大自然碳工厂、制氧车间的重要作用。

Carbon factory and oxygen production workshop of nature

Forest is a green world. As long as there is warm sunshine, photosynthesis will dye the earth green. Forest is rich in biodiversity. It has a complete ecosystem, and is always filled with a strong sense of mystery. Forest is the most active stage of life in the biosphere. Autotrophic and heterotrophic organisms interpret wonderful stories on the stage.

Plants produce a lot of organic matter and oxygen, which are necessary for human survival, through photosynthesis. Only when walking into the forest can we understand animals and plants that live here. We can further experience the importance of forest carbon factory and oxygen production workshop.

大自然的语言

大自然是有语言的,如果静下心来凝听或仔细观察,你就会明白大自然的表达。由于季节的变换,水热条件出现差异,所有植被都会发生季相变化。

因为海拔的差异和水热变化,使得森林里的针叶、阔叶乔木、灌丛、高山草甸和竹林等错杂而生,植被分布表现出显著的海拔变化特征。

伴随着地球四季气温的转换,植物的季相也呈现出不同的色彩。不同的植物在不同的季节都要变换"衣裳"。

"森林萌主"
身边的
那些事
STEP INTO
GIANT PANDA
HABITAT

The language of nature

Nature has language. If you listen carefully or observe carefully, you will understand what nature expresses. Due to the change of seasons and differences in water and heat conditions, all vegetation will undergo seasonal changes.

Difference in altitude leads to the hydrothermal change. That's how conifers, broad-leaved trees, shrubs, alpine meadows and bamboo forests are mixed in the forest, and the vegetation distribution has significant altitude changes.

Change of seasons also makes plants show different colors. Different plants have to change "clothes" in different seasons.

没有花朵和果实的植物

Plants without Flowers and Fruits

生物界的共生典范

地衣是地球上的拓荒者，哪怕在气候严酷、土地贫瘠的地域，也可以看到它的身影。它会分泌地衣酸，是一种通过腐蚀岩石，促进风化过程，使岩石变为土壤的先锋植物。

地衣是藻类和真菌组合共生的复合有机体。地衣共生体的藻细胞能够进行光合作用，为地衣植物体输送养分；菌丝则负责吸收水分和无机盐，为藻细胞的光合作用提供原料，并使藻细胞保持一定湿度，因此，地衣的藻与菌之间构成了一种互惠互利的共生关系。

地衣是大气污染的监测指示植物，地衣类的松萝垂挂在枝丫上，迎风飘舞，既美轮美奂，也为一些动物提供了食物。

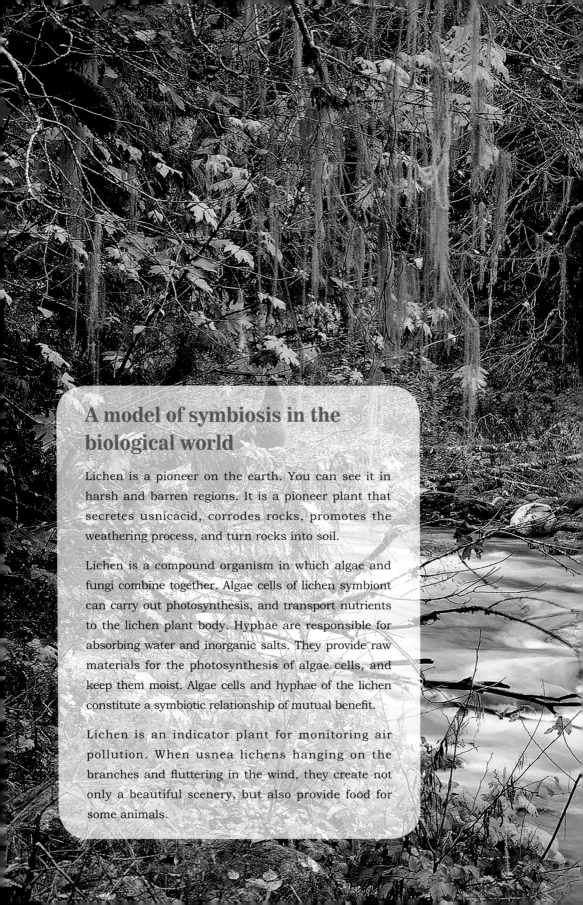

A model of symbiosis in the biological world

Lichen is a pioneer on the earth. You can see it in harsh and barren regions. It is a pioneer plant that secretes usnicacid, corrodes rocks, promotes the weathering process, and turn rocks into soil.

Lichen is a compound organism in which algae and fungi combine together. Algae cells of lichen symbiont can carry out photosynthesis, and transport nutrients to the lichen plant body. Hyphae are responsible for absorbing water and inorganic salts. They provide raw materials for the photosynthesis of algae cells, and keep them moist. Algae cells and hyphae of the lichen constitute a symbiotic relationship of mutual benefit.

Lichen is an indicator plant for monitoring air pollution. When usnea lichens hanging on the branches and fluttering in the wind, they create not only a beautiful scenery, but also provide food for some animals.

大地的绿地毯

  当你步入森林，踩着柔软的苔藓，便如同走上地毯。苔藓是绿色自养性植物，低等种类只有扁平的叶状体，高等类型有假根及类似茎叶的分化，但内部无维管组织。苔藓植物不具硬而粗大的茎和叶，也不开花，没有种子，用孢子繁殖。孢子体简单，寄生于配子体上，多生长于阴湿的环境，如山石、泥土表面及树干或枝条上。它也是植物界的拓荒者之一。苔藓植物都有很强的吸水能力，能起到涵养水分和固土的作用。

Green Carpet of Earth

When you walk into the forest, moss is as soft as a carpet. Moss is a green autotrophic plant. Lower species have only flat fronds, while higher species have false roots and differentiation similar to stems and leaves. However, there is no internal vascular tissue. Bryophytes neither have hard and thick stems and leaves, nor have bloom or seeds. They reproduce through spores. Sporophyte is simple and parasitic on the gametophyte. It grows in humid environment, such as rocks, soil surfaces, tree trunks or branches. It is also one of the pioneers in the plant world. Bryophytes have a strong ability to absorb water and play a role in water conserving and soil consolidation.

自恐龙时代走来的植物

蕨类植物最早出现于泥盆纪,是生物多样性的重要组成部分,对外界自然条件的反应具有高度敏感性。在高等植物中,蕨类植物属于比较低级的一类,有根、茎、叶的分化和较原始的维管组织,不具花,以孢子繁殖,多为草本,少木本。

在野外的林下,蕨类植物长有很多卷曲及密集纤毛的幼叶,叶背面有许多棕色虫卵状孢子囊。蕨类植物的叶形千差万别,有小型叶与大型叶之分。蕨类植物的茎多为根状茎,只有少数种类具有高大直立的地上茎。

Plants since the age of dinosaurs

Ferns appeared in the Devonian and play an important role in biodiversity. They are highly sensitive to natural conditions. Ferns belong to lower class in higher plants category. They have roots, stems, leaves differentiation and primitive vascular tissue, but do not have flowers They reproduce by spores. Most of them are herbs, and few are woody.

In the wild, there are many curly young leaves with dense cilia. There are also many brown worm egg-like sporangia on the back of the leaves. Leaves of ferns are very different. Some are small, and some are large. The stems of ferns are mostly rhizomes. Only a few species have tall and upright above-ground stems.

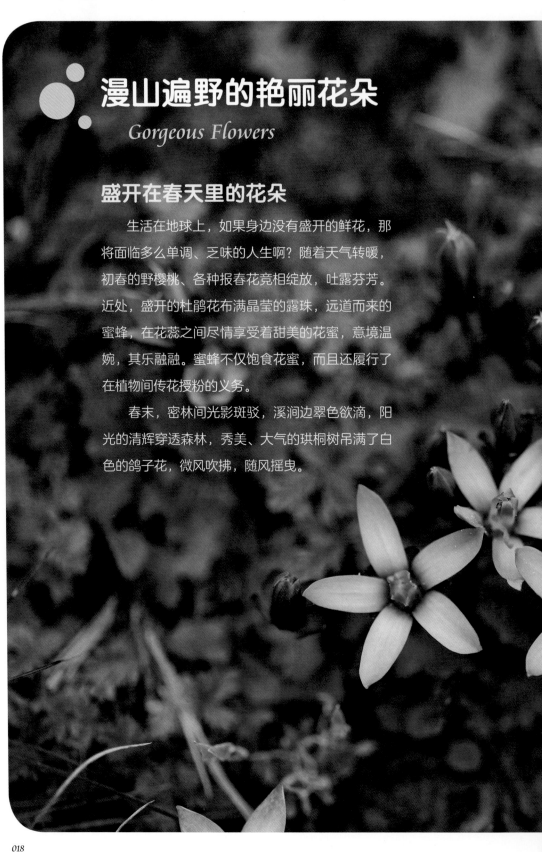

漫山遍野的艳丽花朵
Gorgeous Flowers

盛开在春天里的花朵

生活在地球上，如果身边没有盛开的鲜花，那将面临多么单调、乏味的人生啊？随着天气转暖，初春的野樱桃、各种报春花竞相绽放，吐露芬芳。近处，盛开的杜鹃花布满晶莹的露珠，远道而来的蜜蜂，在花蕊之间尽情享受着甜美的花蜜，意境温婉，其乐融融。蜜蜂不仅饱食花蜜，而且还履行了在植物间传花授粉的义务。

春末，密林间光影斑驳，溪涧边翠色欲滴，阳光的清辉穿透森林，秀美、大气的珙桐树吊满了白色的鸽子花，微风吹拂，随风摇曳。

"森林萌主" 身边的那些事
STEP INTO GIANT PANDA HABITAT

Flowers are in full bloom in spring

Living on the earth, if there is no flower blooming, it is so monotonous and boring. As the weather gets warmer, wild cherry flower and various primroses bloom in early spring and revealing their fragrance. Petals of the rhododendron are covered

with bright drops of water. Bees coming from a far place are enjoying the sweet core. Bees not only eat nectar, but also fulfill the duty of pollination among flowers.

At the end of spring, the light and shadow in the dense forest are dappled. Creeks are dripping with emerald colors. Sunshine penetrates the virgin forest. Beautiful and atmospheric Davidia trees are covered with white pigeon flowers. Breeze blows like vigorous youth.

报春花的春天

　　在冰雪消融、春回大地的季节，除了高大的野樱花在山间盛开，森林的地面也被粉红浸染，那是报春花的杰作。

　　报春花中的多数种类分布于低纬度的高海拔地区，喜生长于温暖湿润和富含腐殖质的土壤环境。

　　早春的山野里，不同种类的报春花次第开放，把森林地表打扮得分外妖娆。报春花的怒放，意味着春天已悄然走进了人们的生活中。

STEP INTO GIANT PANDA HABITAT

Primrose in spring

When snow melts and spring returns to the earth, except for tall wild cherry flowers blooming on the mountains, primroses also color forest with pink.

Most types of primroses are distributed in low-latitude and high-altitude areas. They like to grow in the warm, humid and humus-rich soil.

During early spring, different primroses bloom and dress the forest surface enchantingly beautiful. Primroses blooming means spring has quietly entered people's hearts.

怒放在高山之上的杜鹃花

　　杜鹃花也叫映山红，是中国十大名花之一。

　　由于受海拔和气温的影响，每年五月才是高山杜鹃花盛开的季节。根据生境的不同，杜鹃花植株有高有低，小的伏地而生，高大的乔木杜鹃则巍然挺立，枝干挂满缕缕松萝，随风飘拂，如帘似幕。

　　高山杜鹃是高山木本花卉之王，盛开在山坡上的杜鹃花在流线山体轮廓的映衬下，把一幅幅优美的风景画奉献在我们眼前。它绽放在青山之上，却美艳在人们心中！

Rhododendron(Azalea) on mountains

Rhododendron is one of the top ten famous flowers in China. It is also called Azalea.

　Due to the influence of altitude and temperature, May is the main season for alpine azaleas to bloom. According to the different habitats, azalea plants are tall or short, or even grow on the ground. Branches of tall ones are covered with usnea, which looks like curtains.

　The master rhododendron is the king of alpine woody flowers. Rhododendrons blooming on the slopes presents a beautiful landscape painting in front of us. Standing on the mountain, it leave beautiful scene in people's hearts!

迎风飘舞的鸽子花

中国有一种古老的植物名叫珙桐,也称为鸽子树,已列为国家一级重点保护野生植物。珙桐枝叶繁茂,叶大形圆,于每年四五月份开花,花形似鸽,状如众多白鸽栖息枝头,振翅欲飞,故名"鸽子树"。

珙桐为一千万年前新生代第三纪的孑遗植物,至第四纪冰川时期,大部分地区的珙桐相继灭绝。因为珙桐是在四川中西部山区幸存下来的古老植物,所以被称为"植物活化石"。

Flowers of *Davidia involucrata*

There is an ancient plant called *Davidia involucrata* in China. It is also known as the pigeon tree. *Davidia involucrata* has been listed as a national first-level key protected wild plant. *Davidia involucrata* has luxuriant branches and leaves, with large round leaves and dove-shaped flowers. In April and May, the *Davidia involucrata* flowers bloom. They look like white pigeons are perching on the branches and fluttering their wings. That is why *Davidia involucrata* is also called "pigeon tree".

Davidia involucrata was a relict plant of the Tertiary in the Cenozoic 10 million years ago. During the Quaternary Glacial Period, *Davidia involucrata* was extinct in most areas. Because it was only survived in the central and western mountainous areas, this ancient plant is called "living plant fossil".

优雅的辛夷花

辛夷花又名望春花、紫玉兰,属木兰科的一种。

木兰科植物多为高大乔木,因其姿态清雅、花色动人而深得人们喜爱。木兰科植物虽然种类繁多、资源丰富,但由于屡遭砍伐与破坏,有不少种类已处于濒危状态,如鹅掌楸、西藏含笑、厚朴和峨眉含笑等。辛夷花是其中最具观赏性的一种,色泽鲜艳、芳香浓郁。

Magnolia flower

Magnolia flower, also known as spring flower and purple magnolia, belongs to the Magnoliaceae.

Magnolia plants are mostly tall trees, which are loved by their beautiful appearance, fragrance and pleasant charm. Although there are many kinds of Magnoliaceae plants, but many species are endangered due to immoderate felling and destruction, such as Liriodendron chinensis, Tibetan Michelia, Magnolia officinalis, and Emei Michelia. Magnolia flower is the most ornamental one because of its bright color and heavy scent.

"森林**萌**主"
身边的
那些事
STEP INTO
GIANT PANDA
HABITAT

植物们的生存关系
Coexistence relationship between plants

"适者生存"指引着植物的进化方向

乔木是森林植物的"霸主"。它霸占着顶层,伸展树冠,遮荫蔽日,灌木和草本植物只能见缝插针,"委屈"地生长在狭小的空间里。

大自然的植物在几亿年的生存过程和复杂的生态系统中,一直暗藏着激烈的竞争和适应性变化,并由此不断选择自己的进化方向。

"Survival of the fittest" guides the evolution of plants

In a calm forest, the arbor become the "overlord" of the forest. It will occupy the top layer, stretch the canopy, and shade the sun. Shrubs and herbaceous plants can only grow in a small space furtively.

During hundreds of millions of years fierce competition and adaptive changes, plants constantly choose their own direction of evolution in the complex ecosystem.

植物生存竞争的"角斗场"

　　森林里的一切生物都必须遵循"丛林法则",无论何时何地,都存在着彼此之间的竞争与合作,生存和繁衍的本能始终主导着植物间的故事走向。

　　面对强大的乔木和灌木,草本植物没有强壮的主茎,更没有发达的根系,小草更是处于"被欺凌"的境地。但不屈不挠的草本植物,却巧妙地演化出"早婚早育""多子多福""小个头""短寿命"的本事,成功地找到了用全部能力繁衍后代的生存策略。

　　很多草本植物是一年生,它们充分利用"时间差",早春萌芽,在沐浴阳光的同时快速吸收养分,迅速开花结果,并释放出大量种子,种子快速落地生根,如此循环往复,完成一轮又一轮的生命更替。

"Colosseum" of competition for survival

Everything in the forest must follow the "rules of the jungle." There is competition and cooperation between creatures in everywhere and at any time. Instinct for survival and reproduction always dominates the direction of evolution among plants.

Comparing with powerful trees and shrubs, herbs neither have strong main stem, nor a well-developed root system. Small grasses are in a situation of being bullied. However, herbaceous plants have ingeniously evolved survival strategies, such as "early marriage and early childbirth", "multiple children and more blessings", "small size" and "short lifespan". They use all their capabilities to reproduce offspring.

Many herbaceous plants are annual. They make full use of the "time difference". They germinate in early spring, quickly absorb nutrients while bathing in the sun, quickly bloom and bear fruit, and release a large number of seeds. The seeds quickly fall to the ground and take root. This cycle repeats, and herbaceous plants complete a round of life replacement.

森林里的附生与寄生关系

　　附生植物不跟土壤接触，它们是将自己的根附着在其他树的枝干上生长，利用空气中的水汽、雨露及腐殖质为生，如蕨类、兰科一些植物。它们自身可进行光合作用，不会掠夺被附着植物的营养与水分。

　　寄生植物就不一样，它们会依附在另一植物体中以求得养料，如菟丝子、斛寄生等，它们全靠吸收寄主的养分而存活。

Epiphytes and parasites in the forest

Epiphytes do not contact the soil. Their roots are attached to the branches and trunks of other trees. Plants such as ferns and orchids use the moisture in the air, rain and dew, and humus to make a living. They can carry out photosynthesis by themselves and will not rob the nutrients and water from the plants they are attached.

Parasitic plants are different. They are attached to another plant to obtain nutrients, such as Cuscuta and Dendrobium. They all live by absorbing nutrients from their host.

"森林萌主"身边的那些事
STEP INTO GIANT PANDA HABITAT

山林里藤缠树的大戏

　　诺大的森林里，植物间的生存与竞争无处不在。各种植物为了获取阳光、水分和养分，纷纷演变出各种不同的生存方式，如藤本植物都是缠绕在别的植物之上。

　　在自然界，不管是藤缠树，还是树缠藤，随着绞杀植物的不断长大，它们的侵略性就会逐渐显现，通常会借助寄主树木来支撑自己的躯干，继续向高处攀爬，争夺每一寸阳光就是未来生存的希望！

The drama of vines and trees in the forest

Competitions for survival among plants are everywhere in the forest. In order to obtain sunlight, water and nutrients, various plants have evolved various ways of living. For example, vines twine round other plants.

In nature, no matter it is vines wrapped around trees or trees wrapped around vines, as strangulation plants grow, their aggressiveness will gradually appear. They use host trees to support their torso and climb higher. Every inch of sunshine is the hope for future survival!

自然灾害是植物的一次"凤凰涅槃"

  对于人类来说，谁都不愿意经历自然灾害的伤害和折磨，但对很多植物而言，虽然也是一种挑战，但更是机遇。每一次自然灾害的降临，都会演绎出惊心动魄的故事。

  森林里时常发生山火、山体滑坡、泥石流、山洪等，这些自然灾害会对大地上的生物造成严重伤害，对植物来说，是又一次的重新"洗牌"。但对草本植物而言，却是机遇大于挑战，遭受自然灾害后的大量乔木、灌木很难在短时间内复原，草本植物反倒会快速利用这一生境加速繁殖，从而在残酷的生存竞争中占得一席之地。待木本植物重新占领这个生境时，那又是几十年后的另一次轮回了！

Natural disasters are a "Phoenix Nirvana" of plants

For human beings, no one wants to experience natural disasters. But for many plants, natural disasters are challenges and even opportunities. They always perform thrilling stories.

Wildfires, landslides, mudslides, and flash floods often occur in the forest, which will destroy everything on the ground. For plants, it is another "shuffle". Especially for herb plants, the opportunities outweigh the challenges. A large number of trees and shrubs after natural disasters are difficult to recover in a short period of time. The herb plants will quickly use this habitat to accelerate their reproduction and gain a place in the competition for survival. When woody plants re-occupy this habitat, it will be decades of reincarnation!

"森林萌主"身边的那些事
STEP INTO GIANT PANDA HABITAT

异样的植物生存方式
A different Way of Life for Plants

生活在动植物尸体堆里的生物

在森林底层，生活着一些从腐烂生物体中获取营养物质的生物，它们被称为腐生生物。但凡能将动物尸体和有机物腐烂分解的腐败细菌、大多数霉菌、酵母菌，以及蘑菇、香菇、木耳、银耳、猴头菇、灵芝等，都是典型的腐生生物。由于腐生生物自己不能进行光合作用与制造有机养分，因此属于异养生物中的一类。

大一点的蛆、蚯蚓等动物也属于腐生生物中的一份子，它们都在维持生态平衡中起着重要作用。

Creatures living in dead animals and plants

At the bottom of the forest, creatures get their nourishment from decaying organisms are often called saprophytic organisms. Spoilage bacteria,

most molds, yeasts, mushrooms, shiitake mushrooms, fungus, white fungus, hericium, ganoderma, etc. that decompose dead animals and plants and organic matter are all typical saprophytic organisms. Saprophytic organisms cannot perform photosynthesis and produce organic nutrients by themselves, so they belong to the category of heterotrophic organisms.

Larger maggots, earthworms and other animals are also part of saprophytic organisms, and they all play an important role in maintaining ecological balance.

森林中的"荒漠"

　　森林是由以木本植物为主的、多样化的、完整的植物群落构成，具有森林生态系统的生物多样性，是可为野生动物提供隐蔽、休息、食物的栖息地。

　　因为人类的不断扩张，森林被大量砍伐，导致植物群落受到严重摧毁。虽然人们在砍伐后的林地上栽种树苗，但人工造林的树种极为单一，且成排成行栽种，不具备生态系统多样性的属性。这种通过人工造林营造出的环境，由于缺少丰富的、自然演替的植物群落，不能为动物提供必要的食物与栖息场所，因此被称为"绿色荒漠"。不仅如此，由于品种单一的树木林地，丧失了森林生态平衡和涵养水土的能力，也容易被病害与虫害侵袭或摧毁。

"Desert" in the forest

The forest is composed of woody plants, a diverse and complete plant community. Biodiversity of the forest ecosystem can provide habitat including shelter, rest, and food for wild animals.

Because of the continuous expansion of human beings, forests have been deforested and plant communities have basically been destroyed. People plant saplings artificially on the felled forest land. Trees for artificial afforestation lack diversity and are planted in rows. This kind of planted forest has no abundant natural succession of plant communities and cannot provide food or habitat for animals, so it is called the "green desert". Monoculture woodland not only loses the ability to maintain ecological balance and conserve water and soil, but also can be easily destroyed by diseases and pests.

恐怖的外来入侵物种

生态系统是历经千万年来的竞争、排斥、适应，才形成了相互依赖又互相制约的平衡关系。如果一个外来物种被引入到新的环境中，只要没有与之抗衡或制约它的生物存在，这个新引进的物种就很可能成为真正的入侵者，从而打破原有的平衡，并进而改变或破坏当地的生态环境。外来物种极有可能会破坏原有景观的自然性和完整性，甚至摧毁整个生态系统，危害动植物的多样性。

在我国境内，危害最大的外来物种主要有紫茎泽兰、薇甘菊、水葫芦、福寿螺等。

Terrifying invasive species

The ecosystem has formed a relationship of mutual dependence and restraint after thousands of years of competition, exclusion, and adaptation. When an alien species is introduced into a new environment, if there is no creature that can compete or restrict it, the introduced species may become a real invader by breaking the balance, and changing or destroying the local ecological environment. Alien species may destroy the ecosystem by changing the naturalness and integrity of the landscape, and damage the diversity of plants and animals.

Eupatorium Adenophorum, Mikania Micrantha, Water Hyacinth, and Apple Snail are the most harmful invasive species in China

第二章　**Chapter Two**
Arthropod Zone
节肢动物空间

节肢动物是地球上种类最多、数量最大的一类物种，对生态平衡起着举足轻重的作用，那就让我们从身边最常见的小虫子开始了解它们吧。

Arthropods are the most diverse and abundant species on earth. It plays a pivotal role in ecological balance. Let's start to understand them from the most common species around us

生物圈中不可忽视的节肢动物

人们熟知的蝴蝶、蜘蛛、蚊、蝇都属于节肢动物,是动物界中分布最广、种类最多、数量最大的一类物种。它们的身体由头、胸、腹三分部组成,每一个体节上都长着一对附肢,身体两侧对称,表层全部覆盖着几丁质。在生长过程中会定期蜕皮。它们的生殖方式多样,一般为卵生,多雌雄异体。它们的生活环境极为广泛,水、陆、空都可见它们的身影。

在节肢动物中,蝗虫和白蚁无论在何处,都是不可忽视的群体,它们数量庞大,集群迁移时往往遮天蔽日,可以瞬间改变地表面貌或打破脆弱的生态平衡。

Arthropods that cannot be ignored in the biosphere

The well-known butterflies, spiders, mosquitoes, and flies are all arthropods. Arthropods are the most widely distributed, most diverse, and most numerouscategory in the animal kingdom. The body is composed of three parts: head, chest, and abdomen. Each segment has a pair of appendages. The sides of the body are symmetrical, and the body surface is covered with chitin. During growing, they must shed their skins regularly. They have diverse reproductive methods, including polydioecious and generally oviparous. They live in a wide range of environments, and they can be found in water, land, and air. Locusts and termites are a group of animals that cannot be ignored no matter where they are. They are huge in number, and when they migrate in clusters, they cover the sky and can instantly change the surface of the land or break the fragile ecological balance.

夜晚里闪亮的萤光

萤火虫是典型的环境指示物种，由于受到人类活动的影响，目前在城市周边已很难看到。萤火虫会在夜晚发出一闪一闪的亮光，其中，大部分萤火虫的发光部位都在腹部，而有一种萤火虫的全身都会发出亮光，它就是大场雌光萤。

由于萤火虫体内含有一种被称作虫萤光素酶的化学物质，在与氧气相互作用后，便会产生光亮。在交配季节，大场雌光萤会先用尾部发出亮光吸引雄虫，光点均匀分布在身上，如同在全身挂满了霓虹灯，把夜景装点得如梦似幻。

"森林萌主"
身边的
那些事
STEP INTO
GIANT PANDA
HABITAT

Shining fluorescent light at night

Fireflies are environmental indicators and are rarely seen around cities due to human activities. Fireflies emit a flash of light at night, and most of the light-emitting parts of the fireflies are the abdomen. There is a kind of firefly that emits light all over the body, which is called large-field female firefly.

A chemical substance called luciferase in fireflies interacts with oxygen to produce light. During the mating season, large-field female fireflies will use the tail to attract male insects first. The light spots are evenly distributed on the body as if the whole body is hung with neon lights, and the night sky is decorated like an illusion.

"森林萌主"身边的那些事
STEP INTO GIANT PANDA HABITAT

唱响夏夜交响曲的蟋蟀

夏天的夜晚，鸣叫了一整天的蝉已慢慢收声。然而，不甘寂寞的蟋蟀又唱响了夏夜的交响曲。

蟋蟀俗名蛐蛐、秋虫等，是许多人儿时记忆深刻的小昆虫。蟋蟀常栖于石下、草丛中，多夜出活动，食各种作物、树苗、菜果等。蟋蟀生性孤僻，喜欢独居，它们彼此之间一旦碰到一起，就会发生激烈的争斗。但是，一只雄性蟋蟀却可与多只雌蟋蟀同居。

每到夏夜，蟋蟀就会利用翅膀发声。悠扬的小夜曲，使夜幕的意境更深邃、更撩人。

Crickets play summer night symphonies

In a summer night, cicadas that had been calling for a day began to rest quietly, but the crickets played the nocturne again.

Crickets, also named Ququ or autumn insects in Chinese, are small insects with deep memory in childhood. Crickets often live under rocks and in the grass. Crickets are active at night, eating various crops, saplings, vegetables, and fruits, etc. Crickets are withdrawn by nature and like to live alone. Once they meet each other, they will fight fiercely. However, One male cricket can live with multiple female crickets.

Every summer night, crickets use their wings to make sounds. The melodious serenade makes the mood of the night deeper and more seductive.

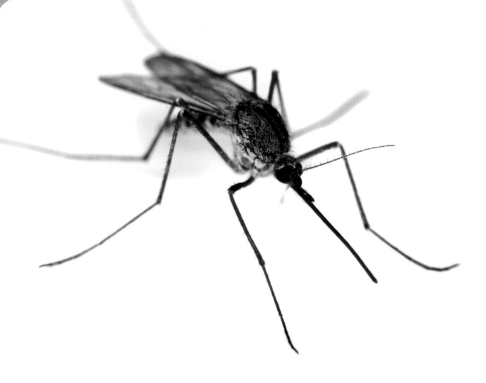

远道而来的吸血者

人们一旦被蚊子叮咬后,皮肤表面就会鼓起红红、痒痒的疹子包,敏感体质的人还会扩散为一大片。很多流行性疾病,如疟疾、黄热病、登革热等,都是通过蚊子的叮咬而传播的。

蚊子属于昆虫。雌性蚊子成熟后,需要大量营养支撑繁殖,因此,雌性蚊子都会对动物的血液情有独钟,是血腥的杀手,而雄性蚊子只吮吸植物的汁液。

我们平常说的"花脚蚊",其实叫白纹伊蚊,它以前主要生活在东南亚气候温暖的地区,随着全球变暖和"花脚蚊"的适应能力逐渐增强,目前已侵入世界各地。

人类和蚊子一直进行着从未间断的战斗,蚊子给人类带来了很多疾病与痛苦,目前还没有找到消灭它的有效办法,这也告诫我们,千万不可小觑外来物种的入侵!

Vampires from afar

After being bitten by a mosquito, a red, itchy rash will appear. Rash in a sensitive body will spread to a large area. Many diseases such as malaria, yellow fever, and dengue fever are spread by mosquitoes.

Mosquitoes are insects. When female mosquitoes mature, they need a lot of nutrients to support reproduction. Therefore, female mosquitoes have a soft spot for animal blood and are bloody killers. Male mosquitoes only suck plant juice.

Aedes albopictus is also known as "decorative pattern foot mosquito". In the past, it mainly lived in warm climates in Southeast Asia. With global warming and the adaptability of the "decorative pattern mosquito", it has invaded all over the world.

Humans and mosquitoes are always fighting. Mosquitoes have brought many diseases and pains to humans. There is no way to deal with them at present. This also reminds people that the invasion of foreign species must not be underestimated!

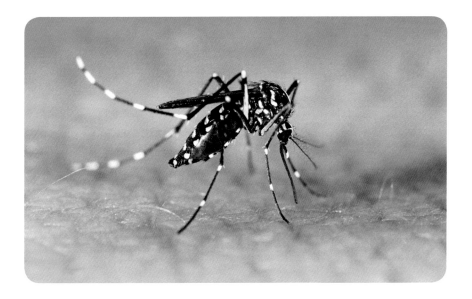

鸣叫不倦的蝉

　　小时候的夏天，人们常常被蝉鸣打断午睡。放暑假的日子，除了捉迷藏之外，还会经常约几个小伙伴一起去村边的林子里捉蝉。蝉的幼虫生活在土壤中，仅靠吮吸植物根部的汁液为生。蝉的幼虫会在土壤里生活几年，甚至十几年，一般为3~5年，个别的长达13~17年。

　　蝉将口器插入树干吮吸植物的汁液，以获取自己生长所需的营养与水分。雄蝉会鸣叫，因腹肌部鼓膜受到振动而发声。为了引诱雌蝉来交配，雄蝉每天都会唱个不停，但雌蝉不能发声，它是"哑巴蝉"。

Cicadas that sing tirelessly

When we were young, we were often interrupted by cicadas during a summer nap. In the summer vacation, except for playing hide and seek game, a few friends often go to the village forest to catch cicadas. The larvae of cicadas live in the soil and suck the sap from plant roots. It will live in the soil for several years (usually 3 to 5 years) or more than ten years (13 to 17 years).

The mouthparts of cicadas are inserted into the trunk of the tree to suck the juice and obtain nutrients and water needed for their growth. Male cicadas make a sound by vibrating tympanic membrane of the abdominal muscles . In order to lure female cicadas to mate, male cicadas sing every day. Female cicadas cannot make a sound. They are "dumb cicada".

会抛大网的蜘蛛侠

　　动物要生存下去，就必须掌握一些攸关自己命运的"绝活"，蜘蛛的"绝活"之一就是网杀。蜘蛛广泛分布于世界各地，属肉食性节肢动物。蜘蛛主要依靠结网获取食物，蜘蛛的种类不同，其结网的形状与结构也不相同。

　　蜘蛛有多种捕杀猎物的招式：第一招是缠绕。蛛丝会死死缠住入网的猎物，同时注射有毒的消化酶，猎物或被毒死，或被蛛丝捆住窒息而死；第二招是突袭。蜘蛛会趁猎物防备不当时进行突袭，通过注射毒液和消化酶致其死亡。有的还会死死咬住猎物不松口，常见于非结网蜘蛛；第三招是诱捕。蜘蛛有时只拉几根丝诱敌深入，待猎物进入可视范围内后，便用蛛丝捕捉。

Spider-Man who can throw a big web

For animals to survive, they must master some "unique skills" that are critical to their destiny. Spiders are widely distributed all over the world and are carnivorous arthropods. Spiders mainly rely on forming webs to obtain food. Different types of spiders have different shapes and structures of their webs.

Spiders have a variety of methods to obtain prey: The first trick, winding. Let the spider silk entangle the prey, and at the same time inject poisonous digestive enzymes. The prey is poisoned or tied up by the spider silk and suffocated to death. The second trick is a surprise attack. When the prey is not ready, the spider attacks the preyand injects venom and digestive enzymes to death. Some non-web spiders will bite their prey. The third track, trap. The spider pulls a few threads. When the prey enters the visible range, spider uses the silk to catch it.

善打群架的马蜂

马蜂也叫胡蜂、黄蜂等,是一种分布广泛、种类繁多的昆虫,也是具有社会性行为的一类昆虫,在受到攻击时会群起反击。

马蜂食性复杂,除了花蜜,成虫还捕食毛毛虫、小青虫等。

马蜂的护幼、护巢习性很强,它们的脾气很暴躁,非常具有攻击性。秋季是马蜂一年中活动最频繁的时节,神经异常敏感,此时的攻击性特别强。

Hornet is good at gang fighting

Wasp also called sand wasp or "yellow" wasps. It is a wide-spread and diverse insect. It is also a type of insect with social behavior. When they are attacked, they will fight back.

The wasp has a complex feeding habit. In addition to nectar, adults also prey on caterpillars and small caterpillars.

The hornet has a strong habit of protecting babies and nests, and their tempers are grumpy and aggressive. Autumn is the most active time for bees. They are extremely sensitive and particularly aggressive at this time.

颠覆人们想象的古老生物

每当太阳西下的黄昏时分,各种蜻蜓都会落在树木的枯枝或玉米叶上。蜻蜓是乡下儿童最容易接近的动物,让人记忆深刻,可如今下雨之前都不太容易看见它们的身影了

蜻蜓是一种原始的昆虫,现在人们看到的蜻蜓,飞行中显得小巧轻盈,但在2.8亿年前,蜻蜓的翅展达到了惊人的710毫米,曾是昆虫中的"巨无霸"。

蜻蜓是食肉性昆虫,它们以苍蝇、蚊子、叶蝉等小昆虫为食。蜻蜓将卵产在水中,幼虫也生活在水中。蜻蜓目昆虫属半变态发育,一生只有卵、稚虫和成虫三个阶段。

蜻蜓的复眼结构和飞行动力特点,成为人类仿生学的热门话题。

Ancient creatures that subvert people's imagination

At dusk, various dragonflies fall on the dead branches of trees and corn leaves. Dragonflies are the most accessible animals for children in the country. They have a deep memory. It is difficult to see them before it rains today.

Dragonfly is a kind of primitive insect. Dragonflies that people see now are small and light in flight. But dragonflies 280 million years ago, with a wingspan of 710 millimeters, was a "Big Mac" among insects.

Dragonflies are carnivorous insects. They feed on small insects such as flies, mosquitoes, and leafhoppers. Dragonflies lay eggs in water, and larvae live in water. Odonata insects belong to hemimetabolous development. They have only 3 stages of egg, larva, and adult in their lifetime.

The compound eye structure and flight dynamics of dragonflies have become a hot topic in human bionics.

飘洒在空中的彩墨

　　夏日,无论你走在乡村的田野,还是步入森林,都会有不知名的彩蝶进入你的眼帘,或在你的周边翩翩起舞,如同入画的彩墨,异常美丽。

　　蝴蝶一生要经历卵、幼虫、蛹和成虫四个发育阶段。幼虫和成虫是它一生中的两个活动期。一般来说,幼虫多为食草性、寡食性,有个别种类食蚜虫。

　　蝴蝶是一种变温动物,体温高低会随着周围环境温度的变化而变化,生命活动会受制于外界温度的支配。

我们平常见到的蝴蝶，多为粉蝶、眼蝶和灰蝶，外观普通、体小、色彩单一，常常在野花、小草中飞舞。还有就是凤蝶、蛱蝶，体型较大，外观艳丽，构型复杂，极具观赏性。其中尤以拟态的枯叶蝶最为特别，为了迷惑天敌而把自己伪装成了一片枯叶。

Colorful ink floating in the air

When we walk in the countryside or into the forest, butterflies will enter your vision and dance around you. Flying butterflies look like a beautiful picture of colorful ink.

A butterfly goes through four developmental stages of egg, larva, pupa, and adult in its life. Larvae and adults are two active periods in their life. Generally speaking, larvae are more herbivorous and oligophagous. There are also a few species of butterflies that feed on aphids.

Butterfly is a kind of temperature-changing (poikilotherm) animal. Body temperature changes with the ambient temperature. Life activities are governed by the outside temperature.

The most common butterfly species around us are white butterflies, satyridbutterflies and Lycaenidae. They are ordinary in appearance, small in size, and single in color. They often fly in the wildflowers and grass. There are also swallowtail butterflies and Vanessa. They are large in size, gorgeous in appearance, complex in configuration, and thus very ornamental. Among them, the mimetic dead leaf butterfly named *Kallima inachus* is very special. In order to confuse the natural enemy, *Kallima inachus* dress up like a dead leaf.

动物界的伪装大师

弱肉强食是"丛林法则"的根基，很多动物因为擅长伪装，而逃过了捕食者的猎杀，伪装便成了一种巧妙而有效的生存手段。伪装的基本原则，是把身体的形态或色彩融入到周围的环境中，从视觉上很难被捕食者发现。而另外一种手法，就是把自己变成凶猛的动物模样，借此吓退捕食者。

许多蝴蝶的幼虫就是与周围的植物混在一起，可以非常隐蔽的藏身在植物丛中，这样的伪装，能有效躲避捕食者，提高生存机率。

多数竹节虫的体色为深褐色，个别呈暗绿色或绿色，整个身体像树枝一样，停留在树上，让猎食动物很难发现。竹节虫多静止不动，有时摇曳身体模仿在风中摆动的树枝，能起到很好的迷惑作用。

Master of camouflage in the animal kingdom

The rule in the forest is the law of jungle. Many animals are good at camouflage to avoid hunting by predators. Disguise has become a clever means of survival. The basic principle

of camouflage is to integrate the body shape or color into the surrounding environment, which is difficult to be seen visually. Another method is to pretend to be a ferocious animal to scare away predators.

Many butterfly larvae mix themselves with local plants and can hide in the vegetation. Such disguise can avoid predators and increase the chance of survival.

Most stick insects are dark brown, and some are dark green or green. Their whole body stays on the tree like a branch, and is almost invisible to predators. Stick insects are mostly static. Sometimes they sway their bodies to imitate the branches that are swaying in the wind as a way of bewilderment.

两栖类动物是一类原始的、变温四足动物，发育周期表现为一个变态过程，幼体以鳃呼吸，成体以肺呼吸，能在陆地生活。两栖动物的变态是一种适应，也反映了由水到陆其主要器官的改变过程。

Amphibians are primitive, temperature-changing, four-legged animals. Their development cycles are metamorphic processes. Larvae use their gills to breathe, while adults use their lungs to breathe, and can live on land. The metamorphosis of amphibians is an adaptation. It also reflects the changing process of major organs from water to land.

Chapter Three
第三章 **From Water to Land**
从水到陆

记忆中的"青蛙王子"

蛙类在食物链中占据着相当重要的地位。在我国,蛙的种类比较丰富,分布较广,特别是大雨之后,山林边、田野里蛙声一片,此起彼伏,它们用蛙鸣烘托出大自然的生机勃勃。

青蛙属于两栖类动物,卵产于水中,先孵化成蝌蚪,再经过生长发育成蛙。蛙类苗条,善于游泳,常以昆虫和其他无脊椎动物为食,多栖息于水边。

青蛙只有雄蛙有气囊,它的鸣叫悠扬、耐听。青蛙用舌头捕食,它的舌头会以极快的弹射速度粘住昆虫,并送进嘴里。

青蛙是与人类接触较多的一类动物,可能大家还清楚记得儿时在小溪边捕捞蝌蚪的情景。青蛙的跳水和游泳姿势被人类模仿,它始终和谐的生活在我们身边!

The "Frog Prince" in Memory

Frogs occupy a very important position in the food chain. China is rich in frogs that have a wide distribution. After heavy rain, the sound of frogs in the fields and mountains makes nature appear vibrant.

Frogs are amphibians that lay eggs in the water, hatch into tadpoles, then grow and develop into frogs. Frogs are slim and good at swimming. They feeds on insects and other invertebrates and inhabit the water.

Only male frog has air sac. Its song is melodious and hard to hear. Frog uses its tongue to prey, and its tongue will stick to the insect at a very fast speed and send it into the mouth.

Frogs are a type of animal that humans have a lot contact with. Maybe everyone still remembers the scene of fishing for tadpoles in the creek as a child. The frog's diving and swimming postures are imitated by humans, and it always lives by our side!

爱打泡泡的斑腿泛树蛙

在山谷的溪涧边或水塘边，人们会偶尔听到近似于"啪、啪、啪"的声音，那就是斑腿泛树蛙的叫声。成对的斑腿泛树蛙会趴在一团泡沫上，腿还不停的来回绞打，那是它们正在交配、产卵。蝌蚪就会从泡沫团中坠落水中，正式开始它的旅程。

斑腿泛树蛙产卵时，雌蛙会先排出一些胶质状物质，并用足部将其打成泡沫，随后才将卵和精液排到泡沫中，并不停的搅拌、绞打泡沫。产卵结束后一周左右，蝌蚪就会从泡沫团中坠落水中，正式开始它一生的旅程。

STEP INTO GIANT PANDA HABITAT

Spot-legged tree frog that loves to play bubbles

In valleys, streams, or ponds, occasionally you will hear frog calls similar to "pop, pop, pop", that is, the call of the spot-legged tree frog. Pairs of spot-legged pan-tree frogs are lying on a ball of foam. Their legs are constantly twisting back and forth. They are mating and laying eggs.

When laying eggs, female frogs will expel some gelatinous material first. Then use their feet to stir the gelatinous substance into a foam. The frog then discharges eggs and semen into the foam, and Constantly stir and beat the foam. About a week after the spawning, the tadpole will fall into the water from the foam mass and officially begin its journey.

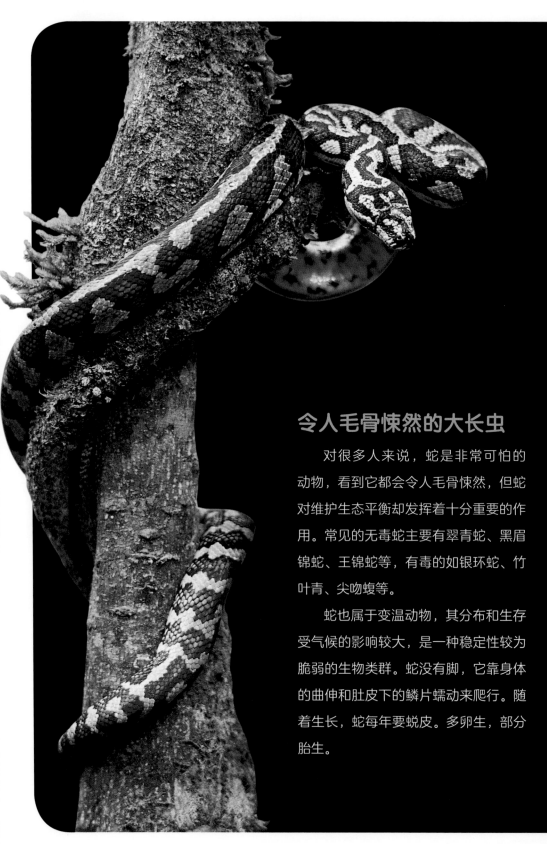

令人毛骨悚然的大长虫

对很多人来说，蛇是非常可怕的动物，看到它都会令人毛骨悚然，但蛇对维护生态平衡却发挥着十分重要的作用。常见的无毒蛇主要有翠青蛇、黑眉锦蛇、王锦蛇等，有毒的如银环蛇、竹叶青、尖吻蝮等。

蛇也属于变温动物，其分布和生存受气候的影响较大，是一种稳定性较为脆弱的生物类群。蛇没有脚，它靠身体的曲伸和肚皮下的鳞片蠕动来爬行。随着生长，蛇每年要蜕皮。多卵生，部分胎生。

蛇的视觉和听觉都不敏感，但它的嗅觉却相当了得！蛇的舌细长、尖端分叉，常伸出口外收集空气中的各种化学物质，借以判断周围的生物环境。蛇还有一手独门绝活，那就是对红外线特别敏感，在捕食和定位小动物时都起着重要作用。蛇是肉食性动物，所吃的动物种类很多，从无脊椎动物到各类脊椎动物均有。

The big creepy worm

For many people, snakes are very scary and creepy animals. Snakes play an important role in maintaining ecological balance. Common non-venomous snakes include green snake, black eyebrow snake, a king snake, etc., and poisonous snakes are silver ring snake, bamboo leaf green, and sharp-nosed vipers.

Snakes are poikilotherm. Their distribution and survival are greatly affected by climate, and they are a group of organisms with relatively fragile stability. The snake has no feet. It crawls by bending its body and squashing scales under its belly. As the snake grows, it sheds its skin every year. Most snakes are oviparous and some of them are ovoviviparous.

The snake's sense of sight and hearing are not sensitive, but its sense of smell is quite good! Snakes' tongues are slender and forked, and they often stick out of their mouths to collect various chemicals in the air to estimate the surrounding biological environment. It also has unique skills and is particularly sensitive to infrared rays, which plays an important role when snakes prey and locate small animals. Snakes are carnivorous animals and they eat many types of animals, from invertebrates to various vertebrates.

蛇中的"淑女"和"猛兽"

蛇作为最常见的爬行动物，总是给人一种冷血、凶猛的印象，但有一种蛇的性格却非常温顺、胆小，基本不主动进行攻击，遇到危险就会立即躲避、逃跑，可谓是蛇类中最为娇羞的淑女了，那就是翠青蛇。

翠青蛇全身翠绿色，无毒，体长大多不到一米，吻端呈圆形，眼睛很大，瞳孔为黑色圆形，常栖息于竹叶或蕨类叶面上，与环境完全融为一体。

但不是所有身体呈翠绿色的蛇类脾气都如此温和，有一种与翠青蛇长相十分相似的蛇，脾气就比较暴躁，那就是竹叶青蛇。如果说翠青蛇是蛇中的"淑女"，那竹叶青蛇就是蛇中的"猛兽"了。竹叶青蛇有毒且具有攻击性，体长不到一米，喜欢栖息于山区树林和溪涧的杂草里，也喜欢吊挂或缠绕在树枝上。

"Ladies" and "Beasts" in snakes

As the most common reptile, snakes always give people a cold-blooded and ferocious impression. However, there is a kind of snake that is very docile and timid, and basically does not take the initiative to attack. It will immediately evade and run when in danger. It can be said that the shyest lady among snakes is the green snake.

Green snakes are emerald green and non-toxic. Their body length is generally less than one meter. Their snouts are round, eyes are large, and pupils are round black. They often live on bamboo leaves or fern leaves and are completely integrated with the environment.

However, not all snakes with an emerald green body have such a mild temper. One type of snake that looks very similar to a green snake has a more grumpy temper, that is, the Green Bamboo Snake. If the green snake is the "lady" among snakes, the green bamboo leaf snake can be described as the "beast". Green Bamboo Snake is venomous and aggressive. It is less than one meter in length. It likes to inhabit mountain woods and stream weeds, and also likes to hang or wrap around branches.

第四章 Bamboo Forest Hermit
Chapter Four
竹林隐士

竹林隐士的生活
The life of the hermit in the bamboo forest

静谧而美丽的熊猫村落

在青藏高原东南的边缘地带,居住着以大熊猫为代表的众多野生动物。该区域山清水秀,终年云雾缭绕,静谧而神秘。大熊猫栖息地周边雪山皑皑,群峰耸立,清澈的溪流从雪山、冰川奔流而下,宛如白色的哈达。大熊猫生活在海拔1500~3500米之间的高山地带,这里的生物多样性极为丰富。可以说,大熊猫居住的村落,是生物圈中最美丽、最有故事的区域。海拔垂直高,可以缓冲气候变暖的影响,不像平原和高原那样会受到剧烈的冲击。大熊猫生活的村落安定、祥和,呈现出一派生机勃勃的繁盛景象。

STEP INTO GIANT PANDA HABITAT

The Quiet and beautiful panda village

The southeastern edge of the Qinghai-Tibet Plateau is the home to many wild animals, including giant pandas. The area is beautiful, quiet, and mysterious all year round. The giant panda habitat is surrounded by snow-capped mountains. Peaks stand tall, and the clear stream rushes down from the snow-capped glaciers, like a white "Hada". Giant pandas live in the forest between 1500-3500 meters above sea level and are rich in biodiversity. It can be said that the village where the giant panda lives are the most beautiful and story-telling area in the biosphere. The altitude is high vertically, which can buffer the effects of climate warming, unlike plains and plateaus that will be severely impacted. The present giant panda village is stable and peaceful, showing a scene of vigor and prosperity.

"熊猫公社"的大食堂

在长期的进化过程中,大熊猫选择了以竹子为食,作为一种远古生物,其物种之所以能生存到今天,毋庸置疑,这是它们最正确的进化方向。在大熊猫的生活环境中,高海拔地区生长着各种冷箭竹,高山和亚高山分布有丰富的拐棍竹、巴山木竹、玉山竹,低海拔地区有白夹竹、箬竹,为它们的生存、繁衍提供了丰富的食源。大熊猫喜食疏林地带的竹子,最为喜爱的是竹笋,它们会根据不同季节、不同竹种萌笋的时间,从低海拔至高海拔"赶笋",最大限度地摄取营养,偶尔也会进食一些野果和小动物。由于竹子是多年生禾本科植物,少有人为破坏和自然灾害,它会生生不息、周而复始的生长,从而为大熊猫的这一天然大食堂,提供了几乎取之不尽、用之不竭的食材。

The canteen of the "Panda Commune"

In the long-term evolution process, giant panda has chosen bamboo as its food. The survival of its species to this day is undoubtedly in the right evolutionary direction. In the living environment of giant pandas, there is a variety of arrow bamboo growing in high altitude areas. The high and sub-alpine areas are rich in *Fargesia robusta, Bashania fargesii, Yushania niitakayamensis. Phyllostachys bissetii and Indocalamus tessellatus* in low altitude areas. Giant pandas like to eat bamboos in sparse forest areas. They love bamboo shoots the most. They will "follow bamboo shoots" from low altitude to high altitude according to different seasons and the time of different bamboo species to maximize nutrition. Occasionally they eat some wild fruits and small animals. In the natural canteen, it is almost inexhaustible for giant pandas. Because bamboo is a perennial gramineous plant. There is no human damage and natural disasters. It will grow endlessly and repeatedly.

"吃相难看"的大熊猫

大熊猫吃竹子，首先要一看二闻，然后才选取可口的进食，根本不需要试吃。

大熊猫最喜欢的一口是竹笋，就如同四川人喜欢吃回锅肉一样。它们剥掉笋壳与去除竹竿表皮的技巧相当高超，完全可以注册"专利"！这剥皮的功夫，主要源于大熊猫的前肢多出的一个伪拇指和牙齿密切配合、协同的功劳。它们还时常切换左右牙床，可谓相当讲究。

最让人难以接受的是大熊猫的"吃相",对食物喜爱的行为非常外露,丝毫不加掩饰,看起来吃得特别香,再配上"吧唧……吧唧……"咀嚼的声音,让人不禁口水直流!不仅"吃相难看",而且相当"虐人"!

Giant pandas with an inelegant way of eating

To eat bamboo, a giant panda must first look at it and smell it, and then choose delicious part. There is no need to try first.

Giant pandas' favorite are bamboo shoots, just like twice-cooked pork that Sichuanese like. Giant Pandas have excellent skills in peeling and removing the skin of bamboo shoots and can be registered for "patents"! This peeling effort is mainly since the giant panda's forelimbs have a pseudo-thumb and teeth that closely cooperate. They often switch between the left and right gums, which is quite particular.

The most unbearable thing is the giant panda's "eating appearance". The behavior of loving food is very exposed. It seems to eat very fragrantly. When coupled with the sound of chewing, it makes people salivate! !Binge-watching giant panda devouring food, which makes us wonder: were they taught proper table manners!

The simple life has deep meaning.

In addition to foraging for bamboo, giant pandas generally do not active too much in their daily lives. Energy conservation is the top priority. Giant pandas in the wild will sit among the bamboos, eat without moving their nests, and change places after eating. Because bamboo, a high-fiber food, has low nutrition, unnecessary energy consumption must be reduced. Therefore, people who watch giant pandas always feel that giant pandas are lazy, but they do not know that giant pandas are energy-saving experts.

小日子大道理

大熊猫除了觅食竹子外,在日常生活中一般不会过多活动,节能是第一位的。野外的大熊猫会安坐在丛竹中不挪窝地吃,直到吃完了才换地方。因为竹子是高纤维食物,没有什么营养,必须减少不必要的能量消耗。所以,人们总觉得大熊猫性情懒惰,则不知它是节能高手。

大竹熊不冬眠

在自然界中,很多动物为了度过缺少食物的严酷寒冬,都会找个相对安全和温暖的环境进行冬眠。那大熊猫为什么不冬眠呢?一是以竹子为主食的大熊猫,很难积累和储存大量脂肪撑过漫长的冬眠;二是在深山老林中,虽然各种竹子被冬季的积雪覆盖,但竹叶仍然翠绿,能够完全满足大熊猫的营养需求。

The big bamboo bear does not hibernate

In nature, many animals find a relatively safe and warm environment to hibernate in order to live through the harsh winter when food is lacking. Why don't giant pandas hibernate? One is that it is difficult for giant pandas whose main food is bamboo to accumulate and store a large amount of fat to support their long hibernation. Second, in the winter in deep mountain and forests, although various bamboos are covered by snow, bamboo leaves are still verdant and can satisfy nutritional needs of pandas.

高高大树之上的比武招亲

每年的春季都是一个重要的时节,爱情、竞争、繁衍都在此刻交织在一起,寻偶成了这段时间的首要任务。成年雌性大熊猫发出招郎信号后,周边的成年雄性大熊猫们便会蜂拥而至,此时必须比武过招,只有胜者方能获得交配权。这场残酷的角斗,打得天昏地暗,除了地面上的较量,还常常搏斗到树顶,这个"比武招亲"的过程,往往浸染着刺激而且浪漫的色彩!

春天来临的时候,如果你的运气足够好,在四川卧龙的"五一棚"观测站和佛坪的"三官庙"管护站,或许就能有幸看到大熊猫迎娶新娘的结婚仪式!

A fight for a bride on a tall tree

Spring is an important season every year. Love, competition, and reproduction are intertwined. Finding a spouse becomes the top priority during this time. When adult female giant panda shows signal of courtship , surrounding adult male giant pandas will flock to her. At this time, they have to fight for a bride before the winner get the right to mate. This cruel battle was fought so dimly, not only on the ground but often on the top of trees. This cruel battle was extremely fierce. This "marriage" process is tainted with excitement and romance!

When spring comes, if you are so lucky, you may observe the wedding ceremony of giant pandas at the "Wuyipeng" observation station in Wolong and the "Sanguanmiao" management station in Foping!

"森林萌主"身边的那些事
STEP INTO GIANT PANDA HABITAT

博爱满满的大熊猫

　　大熊猫不仅外貌喜人、动作可爱，而且它还是一个极具博爱精神的"慈善家"。圈养大熊猫种群能够有今天的昌盛，全得归功于对大熊猫母兽具有伟大母爱的发现！直到20世纪90年代初，成都动物园的科技人员才攻破了大熊猫的繁育难题，其关键之处，就在于发现大熊猫母兽能够接受"义子"。在圈养大熊猫的繁殖过程中，经常会出现大熊猫的弃子行为，采用"奶妈"的方式，让其他母兽代为哺乳"义子"，让幼子吃上母乳，大大提高了大熊猫幼仔的成活率。大熊猫母兽的这种"博爱"，帮助科技人员取得了历史性突破，"大熊猫双胞胎育幼研究"获得国家科技进步二等奖！为后来大熊猫圈养种群数量的增长铺垫了坚实的基础。

Giant panda are full of fraternity

Giant pandas are not only cute in appearance and actions, but also actually a veritable "philanthropist". The prosperity of the captive giant panda population today can be attributed to the discovery of great maternal love for giant panda mothers! It was not until early 1990s that scientists and technicians at Chengdu Zoo broke through the breeding problem. The key point is that mothers can accept "Adoptive son", the youngest can survive by taking breast milk. In the breeding process of captive giant pandas, the behavior of abandoning children often occurs, and the "nanny" method is adopted to solve the problem of breastfeeding "Adoptive son" by mother of giant pandas. The "fraternity" of the giant panda mother helped scientific and technological personnel to make a historic breakthrough, and the "Research on Giant Panda Twin Breeding" won the second prize of the National Science and Technology Progress Award! It laid a solid foundation for the future growth of the captive population of giant pandas.

爆啃老山羊

　　大熊猫主要以竹子为食，但还是要吃一些野果和小动物。大熊猫的祖先是由食肉动物进化而来，虽然它们的消化系统适应了长期食竹的需要，但偶尔遇到其他动物尸骨和能够逮到的动物，大熊猫还是乐于"开荤"的。特别是冬季，大熊猫会下到低海拔地区活动，遇到温顺的山羊，它们就会兽性大发。吃了山羊不说，有时还要借宿农家的猪圈大睡一觉，也许大熊猫认为这些都是大自然的馈赠……

　　由于人类的急剧增长，人们不断拓荒种地，扩张生活范围，侵占了原本属于大熊猫们的生活领地。一旦进入漫长的冬季，当大部分可食竹被厚厚的积雪覆盖的时候，大熊猫也会摇动心思去逛逛附近的村庄，大熊猫这种偶尔"还乡"的举动，还是可以理解的吧。

Devour the goat ferociously

Giant pandas mainly feed on bamboo, but they still eat some wild fruits and small animals. Giant pandas evolved from carnivores. Although their digestive system has adapted to eating bamboo for a long time, they occasionally encounter other animal corpse or living animals that can be caught. They are happy to resume a meat diet. Especially in winter, giant pandas will go down to low-altitude areas for food. When they encounter a docile goat, they will of course become wild. After giant pandas eat the goats, sometimes they will take a good night's sleep in the pigpen of the farmhouse. Giant pandas take this for granted.

Due to the rapid growth of human beings, many farmers have opened up wasteland and expanded their living areas, and occupied the original living areas of giant pandas. During long winter, when most of the edible bamboos are covered by thick snow, they will also be tempted to visit the village for food. It is understandable that giant pandas occasionally "return home".

"森林萌主"
身边的
那些事
STEP INTO
GIANT PANDA
HABITAT

大熊猫的艰难经历
Difficult experience of giant pandas

走出第四纪冰川期和地质灾害

在地质历史上,地球曾经多次出现过气候寒冷的大规模冰川活动时期,人们称之为冰河期。对大熊猫影响最大的一次,是第四纪冰川期的来临,当时地球气温平均下降了10℃~15℃,三分之一的大陆被冰雪覆盖。

在第四纪冰川期时,不仅气候发生了剧烈起伏,地球的构造运动也相当活跃,陆地上新的造山运动以喜马拉雅山等地最为剧烈。地震和火山是新构造运动的表现形式,对生物影响巨大,大熊猫也不例外。

第四纪冰川期时,大陆冰盖向南扩展,大熊猫等动植物也随之向南移动。至第四纪冰川后期,大型陆生哺乳动物发生过大规模绝灭,大熊猫却有幸在自然环境剧烈波动的过程中生存了下来。一个物种的首要任务就是努力地活着,然后才是延续,正如大熊猫。

The dark age of giant pandas

Giant panda is a species endemic to China and is the national treasure of China. it is honored as the world peace ambassador. Do you know that Giant pandas were once the target of poaching by western explorers?

In 1840, the first Opium War blasted China through. Since then, Western missionaries have continued to enter China. In March 1869, the French naturalist Father David was on an investigation in Muping, Baoxing, Sichuan, and learned about the existence of giant pandas. In early May of that year, with the help of local hunters, a live giant panda was captured and died afterwards. David made it into a specimen and put it on display at the National Museum in Paris, France, which caused a sensation in the Western world. Since then, a steady stream of Western explorers has continued to come to China, including two sons of former US President Franklin D. Roosevelt, who have traveled thousands of miles in search of giant pandas.

An American from William Harkness came to China to look for giant pandas and died of illness. His widow Ruth returned to China, where she captured a baby panda. Ruth named it "Su Lin" and later smuggled it back to the United States. After Ruth smuggled "Su Lin" to the United States, it sparked another frenzy of theft. From 1936 to 1941, United States alone captured 9 pandas from China. That is why it is said that Western powers have hurt giant pandas in history.

大熊猫的黑暗时代

　　大熊猫是中国的特有物种，也是中国的"国宝"，被誉为"世界和平大使"。你可知道大熊猫曾是西方探险家偷猎的目标吗？

　　1840年，第一次鸦片战争轰开了中国国门，此后，不断有西方传教士进入中国。1869年3月，法国博物学家戴维神父在四川省宝兴县穆坪镇考察时，知道了大熊猫的存在。当年5月初，他在当地猎人的帮助下，捕获了一只活体大熊猫，但后来死亡。戴维将它做成标本，送到法国巴黎国家博物馆展出，顿时在西方世界引起了巨大轰动。此后，不断有西方探险家来到中国，连美国前总统罗斯福的两个儿子都不远万里进入中国寻找大熊猫的踪迹。

　　有个名叫威廉·哈克利斯的美国人来到中国寻找大熊猫，后因病去世，他的遗孀露丝又来中国，并抓获了一只幼年大熊猫。露丝为它取名为"苏琳"，后来偷运回美国。露丝将"苏琳"偷运到美国后，又引发了新一轮的偷猎狂潮。从1936~1941年，仅美国就从中国捉走9只大熊猫。所以说，西方列强曾经伤害过大熊猫。

竹子开花的大饥荒岁月

20世纪70年代末,大熊猫主要分布在岷山、邛崃山、大相岭区域。由于竹子大面积开花、死亡,导致以竹子为生的大熊猫遭遇了罕见的大饥荒。1976年,在四川和甘肃的大熊猫保护区内,仅几个月时间内就发现了138具大熊猫尸体。因病饿而死的大熊猫,其皮毛粗糙无光,体重只有正常时的一半。经解剖发现,大熊猫内脏几乎无脂肪储存,肠胃中的食物残渣很少,普遍存在大量蛔虫,少则几百条,多则数千条。20世纪80年代就有200多只大熊猫因病饿而死,致使大熊猫物种处于濒临灭绝的危险境地。

竹子开花是一种生理现象，具有周期性，一般发生在天气长期干旱的环境条件下。竹子大面积开花后便会逐渐枯死，新的竹子需要几年的成长时间，有些种类甚至需要十年以上才能恢复。竹林的大面积死亡，给生态环境造成了严重破坏，致使大熊猫因食物短缺而走向死亡。

The Great Famine Years during Bamboo Blossom

In the late 1970s, giant pandas were mainly distributed in Minshan, Qionglaishan and Daxiangling areas. Large areas of bamboo bloom and die, forcing giant pandas who depend on bamboo to suffer a great famine. In 1976, 138 giant panda bodies were found in the reserves of Sichuan and Gansu within a few months. Giant pandas that died of illness and starvation had rough and dull fur and only half of its normal weight. The autopsy revealed that there was almost no fat storage in their internal organs, and there were few food residues in their stomach and intestines. There were a large number of roundworms, ranging from a few hundred to as many as thousands. More than 200 giant pandas died of starvation in the 1980s, which made giant panda species be in danger of extinction.

Flowering of bamboo is a physiological phenomenon with periodicity, which generally occurs under long-term dry weather conditions. After a large area of bamboo blooms, bamboo gradually dried up. It takes several years for new bamboo to grow, and some species take more than ten years to recover. The large-scale death of bamboo forests has caused serious damage to the ecological balance of environment, resulting in the death of giant pandas due to lack of food.

"森林萌主"身边的那些事
STEP INTO GIANT PANDA HABITAT

大熊猫"回家"之路

　　大熊猫是我国特有珍稀物种,在中国政府的不懈努力下,就地保护与异地保护都取得了可喜的重要进展,种群数量持续稳定增长。如今,随着天然林保护工程取得显著成效,以及圈养大熊猫数量大幅增加的情况下,让大熊猫"回归家园"理应提上议事日程。

　　大熊猫"祥祥"经过野化培训后,于2003年进行了圈养大熊猫历史上的第一次野外放养,后来又进行了几次野放尝试。2012年放归的"淘淘",已经完全适应了野外独立生活,真正实现了大熊猫"回家"的目标。

　　大熊猫的"回家"之路可谓异常艰难,要把人工圈养的大熊猫放归野外,野化训练是一个漫长而复杂的过程。除了需要恢复因人工饲养丢失的行为,其野外觅食、寻找水源、躲避天敌等生存本领都需要强化。

此外，野放环境也是一个不可控的变量，所以还须随时监测大熊猫身处野外的全部生活情况，包括身体状况是否健康等。目前，中国大熊猫保护研究中心大熊猫野放工作效果显著，离完全铺设大熊猫的"回家"之路已经不远，可以说，大熊猫的保护事业已经迈入新纪元。

The road back home for giant pandas

Giant pandas are unique and rare species in China. With unremitting efforts of the Chinese government, significant progress has been made in both in-situ and ex-situ conservation. Population of giant pandas has grown steadily. Nowadays, as the natural forest protection project has achieved remarkable results and the number of giant pandas has also increased significantly, the "return of giant pandas" should be on the agenda.

After wild training, the giant panda "Xiang Xiang" was released to nature for the first time in 2003. Later, there were several wild release attempts. In 2012, the giant panda "Taotao" was released. It has fully adapted to an independent life in the wild and has truly achieved the goal of giant pandas to "go home".

The giant panda's journey home is extremely difficult. To release captive giant pandas into the wild, wild training is a long and complicated process. Except for restoring lost behaviors from artificial feeding, foraging in the wild, searching for water sources, and avoiding natural enemies are also need to be strengthened.

The wild environment is also an uncontrollable variable. It is necessary to monitor all living conditions of the giant pandas in the wild, including their health. At present, the giant panda release work of the China Conservation and Research Center for Giant Panda has achieved remarkable results, and it is not far from fully laying the road for giant pandas to "home". Giant panda protection has entered a new era.

大熊猫的烦恼与期待

　　由于人类的经济与生产活动日益频繁，不仅压缩了大熊猫的栖息地，而且还把大熊猫的几个族群切割成了彼此孤立的状态，各个山系之间相互通婚的难度很大，由此产生了恋爱难、结婚难、近亲繁殖系数增大等问题，遗传活力明显降低，严重影响到大熊猫家族的繁衍生息。我们应当主动为大熊猫架起能让它们相互沟通的桥梁、打通走廊带、退耕还林，重现大熊猫自然栖息地的生态环境，不久的将来，大熊猫一定会家族兴旺。

　　由于人类在森林里的活动日益增多，采笋、挖取中草药等生产活动频繁，这些行为不仅干扰了大熊猫的宁静生活，而且人类、家畜和家犬还带入了大量的寄生虫和病菌，这些因素同样严重威胁到大熊猫的生存。

　　现在虽然没有了大规模的猎杀行为，但下套索捕猎野生动物的现象还时有发生，那些无处不在的套索，时时刻刻威胁着野生动物的生命，只是不知何时，那些美丽的生灵就会经过极其痛苦的挣扎而无声无息地离开这个世界。

Troubles and expectations of giant pandas

Due to the increasing frequency of human economic and production activities, not only the habitat of giant pandas has been compressed, but also several populations of giant pandas have been cut into isolation. At present, it is difficult for several populations to intermarry with each other, which has caused problems such as difficult love, a difficult marriage, increased coefficient of close relatives, and decreased genetic vitality, It seriously affects the revitalization of giant panda families. They expect people to build bridges, make corridors, return farmland to forests, and restore the ecological environment of their natural habitat for them. In near future, giant pandas are expected to have flourished families.

Human activities in the forest are increasing. Frequent production activities such as harvesting bamboo shoots and digging Chinese medicines not only interfere with the peaceful life of giant pandas but also bring in a large number of parasites and bacteria to humans, domestic animals, and domestic dogs . The same serious threat to the survival of giant pandas.

Although there is no large-scale hunting behavior, the phenomenon of lasso hunting wild animals still happens time to time. Traps threaten lives of wild animals at all times. We have no idea when those beautiful creatures will go through extremely painful struggles and leave the world silently.

第五章 Exotic Animals
Chapter Five
珍禽异兽

莺飞凤舞的森林空间

Auspicious (Harmonious) Forest

看喙识鸟

鸟喙的主要功能是用于进食，但也具备爪的作用，搬运东西时可以用喙叼取，危险来临时可作为打斗的武器，另外还可以用来梳理羽毛。

鸟喙的形状与食性有着密切的关系，其一般规律是：细而弯曲的是采食花粉和小虫的鸟，如蜂鸟、太阳鸟等；尖直的是食虫鸟，如山雀、绣眼、啄木鸟等；直短的圆锥形喙，多是食谷物、种子的鸟，如文鸟、相思鸟等；尖端带弯钩的主要是食肉鸟，如金雕、草鸮、秃鹫等；长而直的主要是以鱼和虾为食的鸟，如翠鸟、鹭鸶、黑颈鹤等；扁平的大多为杂食鸟，如鸳鸯、绿头鸭等。

Identifying Birds by the beak

A bird's beak is used for eating, sometimes it can be used as a claw. It can be grabbed when carrying things, fights when danger comes, and it can also be used to comb feathers.

The shape of a bird's beak is closely related to its diet. In general, thin and curved birds that feed on pollen and small insects, such as hummingbirds and sunbirds; The birds with straight-pointed beak are

insectivorous; chickadees, white eyes, and woodpeckers are all insectivorous birds; Mostly birds with straight and short conical beak eat grain and seeds, such as Munia and Acacia; the birds with curved hooks at the end are mainly carnivorous birds, such as golden eagles, grass owls, and vultures; long, straight beaks mainly feed on fish and shrimp, Kingfishers, herons, black-necked cranes are represented; The birds with flat beaks are mostly omnivorous, such as crows, mandarin ducks, and mallards.

"森林萌主"
身边的那些事
STEP INTO GIANT PANDA HABITAT

大名鼎鼎的森林"医生"

在大树林里,时不时会听到啄木鸟用喙敲击树干的声音,它这是在用那坚硬的喙不断敲碎木质层,以便活捉隐藏在树干深处那些肉嘟嘟的虫子。因为啄木鸟常年在森林里捉虫子,所以人们又把啄木鸟称为"森林医生"。其实,啄木鸟很多时候在枯木上寻找虫子,应该是它生存的本能而已。

人们看到啄木鸟孜孜不倦地用力敲击树干,自然而然就会想,它这样做会不会伤及颈部和大脑呢?请不要担心,啄木鸟的头部骨骼结构疏松,里面具有较大的空间,头颅内部不仅附有一层坚韧的外脑膜,脑髓与脑膜之间还有一条狭小的空隙,其中充满液体,这样便会起到有效的减震作用。严格来说,啄木鸟在林间不辞劳苦地捉虫,应该是生存的本能行为,间接起到了帮助树木除掉虫患的作用,人们称它为"森林医生",也算是实至名归吧!

The famous forest "doctor"

The sound of woodpeckers knocking on tree trunks is often heard in the big woods, using its hard beak to break the woody layer uninterruptedly and catch fleshy bugs alive in the depths. Since woodpeckers catch insects in the forest all year round, they won the fame of the forest "doctor". In fact, woodpeckers often hunt for insects on dead wood,

People see woodpeckers working hard on the tree trunks, and they naturally wonder if they hurt their necks and brains. Please don't worry, the bone structure of the woodpecker's head is loose and there is a large space inside. The inner skull is attached with a tough outer meninx There is a small gap between the brain and meninges, which is filled with liquid, which will have an effective damping function. Strictly speaking, woodpeckers working tirelessly to catch insects in the forest should be an instinctive behavior for survival, which indirectly helps the trees to get rid of insects. So it deserves the fame of forest doctor.

背有一身故事的鸟

清明节后,正是播种的好季节,布谷鸟也在此刻不停地发出"布谷……布谷……"的叫声,所以,人们认为布谷鸟是天地间极具灵气的一种"报信鸟"。

关于布谷鸟,曾在民间流传着一个凄凉的传说:古时候,有一个名叫博古的孩子,他的继母悄悄将黄豆炒熟,然后让他把熟豆种到山上的田里,然后用此阴招陷害博古。后来,含冤离世的博古把自己变成了一只布谷鸟,见到继母就会发出凄惨的叫声:"光种不出,害我博古!"继母听到后,越发觉得恐惧,而后一病不起,呜呼哀哉!

布谷鸟还有一个我们大家都熟悉的名字——杜鹃鸟。"杜鹃啼血"的典故同样充满了凄凉、哀婉的气氛,说的也是这种鸟。

另外,布谷鸟还有一个让人十分不解的行为,有少数布谷鸟会把卵产在食性相同的鸟巢里,让"奶妈"辛勤孵化和养育它的幼鸟。人们觉得这种做法太残酷,但其中肯定有生物学方面的原由,让我们今后一起研究,争取早日揭开这个谜底。

A bird with stories

After the Tomb Sweeping Day, it is the season for sowing, and cuckoos keep calling "cuckoo cuckoo, sow grain quickly". Thus people believe that the cuckoo is an aura of "announcement bird" in the world.

There is a sad story about this bird. In ancient times, the stepmother fry soybeans let the child named Bogu plant them in the fields on the mountain, framed and forced

Bogu to death. Later, Bogu turned into a cuckoo, and when he saw his stepmother, he would make a miserable cry: Soybeans, no seeding, flood Bogu with pain! After hearing this, the stepmother felt more and more frightened and died later with fear and regret.

In addition, cuckoo birds also have a behavior making people confused. A few cuckoo birds lay their eggs in other birds' nests with the same feeding habits, and let them do hard nurses work to incubate and raise their young birds. This is so cruel. There must be a biological reason for this. Let us study together in the future and try to solve this mystery as soon as possible.

雄性雉鸡体色艳丽的理由

雉鸡类是自然界中的一个重要动物类群,如绿尾虹雉、褐马鸡、四川山鹧鸪等,都是非常珍稀的雉鸡类品种。

大自然中的雄性雉鸡,外观都十分艳丽,比雌性雉鸡要漂亮很多,这是为什么呢?其实这也是生物自然选择的结果,雄性雉鸡把自己装扮得能够吸引、取悦对方,最终目的是为了获取雌性的选择与宠幸,毕竟在大自然中,雌性在生殖繁衍上的地位要高于雄性,雌性掌握着主动权和选择雄性的权利。

雄性凭借美丽的外表、强壮的身体展示自己的优良基因,从而得到雌性的青睐,这样更利于让优良的基因得到延续。

The reason of male pheasant's beautiful color

Pheasants are an important group of animals in nature. Such as Chinese monals, brown eared-pheasants, and Sichuan partridges, they are all very rare pheasants.

The appearance of male pheasants in nature is very beautiful, even more beautiful than females. However, this is the result of natural selection. Dressing oneself up can attract and please the partner. The ultimate goal is to obtain partners' choice and favor. After all, females have a higher status over males in reproduction in nature, and females hold the initiative and the right to choose males.

Males rely on their beautiful appearance and strong body to display excellent genes, so as to get the favor and choice of females, so that their excellent genes can be maintained

Spring River Plumbing Duck Prophet

Natatorial birds are an important group of bird species. Common ones include swans, bar-headed geese, mallards, and mandarin ducks. They like to live on the water, with their feet stretched back, webbed toes, and a flat mouth. They are good at swimming, diving, and obtaining food in the water. They are amphibious and good at flying..

Swimming and diving are the main activities of natatorial birds in the water. Their feathers are thick and dense, and their down feathers are well developed.

In order to prevent the feathers from being soaked by water, natatorial birds have developed preen glands to secrete oil, and they use their beaks to smear the feathers for waterproofing. Some natatorial birds have underdeveloped tail fat glands and need to dry their feathers to ensure flight normally. The beaks of swimming birds are mostly flat, eating aquatic plants, fish, and invertebrates. Natatorial birds live in groups and often move in groups.

According to the change of seasons and temperature, natatorial birds will choose foraging spots. And migrate according to the air and water temperature, so it is said that "Spring River plumbing duck prophet".

Patient snake eater

The heron is a common bird, also known as the grey heron. Feeds on small animals such as fish, shrimp, frogs, and insects. It is often seen in rivers, lakes, swamps and other environments. A couple of herons breed at least twice a year, with three to four herons per litter.

The heron is different from the egret in the predation method of wading and walking. It is used to foraging alone and often stands still for a long time. It can wait for several hours in one place, or even half a day without changing positions, waiting patiently for its prey. Therefore, the heron has the nickname "Always waiting".

Heron has a mixed diet. In addition to eating fish and frogs, it will not let go of snakes which is a nourishing meal! It will quickly bite the crucial point of the snake with its long mouth, and then repeatedly beat it up and down. The snake will be swallowed while still moving! Just like eating the "spicy strips", when you are hungry.

把蛇当辣条吃的"老等"

苍鹭也称灰鹭,是一种常见鸟类,以鱼、虾、蛙类、昆虫等小动物为食,常在河边、湖泊、沼泽等环境中看到它们的身影。一对苍鹭夫妇每年至少繁殖两次,每窝三至四只。

苍鹭与白鹭采用涉水行走的捕食方法不同,它习惯于独自觅食,常常一动不动地站立很久,它们可以在一个地方待上几个小时,甚至半天都不挪动位置,非常有耐心的等待猎物的到来,因此,人们给苍鹭取了一个"老等"的外号。

苍鹭食性较杂,除常吃鱼、蛙外,遇到蛇也不会放过,那可是一顿滋补大餐啊!它会用长嘴迅速咬住蛇的七寸,然后反复摔打,就算蛇还在动也会被它吞掉!饥肠辘辘的时候,"老等"一定觉得送上门的"辣条"特别香。

起舞在高原的鹤

黑颈鹤是世界上唯一在高原上生长、繁殖的大型涉禽。黑颈鹤的食性很杂,除了植物的叶、根茎、水藻等之外,也食昆虫、鱼、蛙等。它们在觅食的时候很少用脚,而是用尖嘴在浅水中捕捉动物或从泥土中掘取食物。

越冬期间很少有大的群体,一般是以3到5只的小群分散觅食。黑颈鹤时常高昂头颈,边鸣叫,边起舞,常常引起其他同伴的共鸣。

黑颈鹤春季迁到繁殖地,秋季到达越冬地。每年秋季,黑颈鹤都会带着亚成体鸟结群南迁,从四川的若尔盖、松潘出发,迁移至贵州、云南、西藏等越冬地的沼泽及湿地。夏季的高原湿地,水草丰美,食物充足,地广人稀,最适合养育下一代,因此,川西北的若尔盖高原湿地,一直是黑颈鹤们的重要繁殖基地。

Cranes dancing on the plateau

The black-necked crane is the only large wading bird that grows and breeds on the plateau. Black-necked cranes are Omnivores, feed on food like plant leaves, rhizomes, algae, etc. They also eat insects, fish, and frogs. They seldom use their feet when searching for food, but use sharp beaks to catch animals in shallow water or dig food from the soil.

There are few large groups during the winter, usually small groups of 3 to 5 scattered for food. Black-necked cranes often raise their heads and necks, chirping and dancing, which often resonates with other cranes.

They move to the breeding ground in spring and reach the wintering ground in autumn. Every autumn, black-necked cranes migrate south with sub-adult birds, from Ruoergai and Songpan to the swamps and wetlands of Guizhou, Yunnan, Tibet and other places where they are wintering. The plateau wetlands in summer are rich in water, grass, abundant food, and sparsely populated. They are most suitable for nurturing the next generation. Therefore, the Ruoergai plateau wetlands in northwest Sichuan have always been an important breeding ground for black-necked cranes.

夜幕下的迷藏

　　森林里的夜晚，阴森可怕，没有点功夫的动物们都会隐蔽起来，陡峭的岩壁上、树的顶端、树洞、岩洞、茂密的灌木丛都是藏身之处。夜色中的森林，漆黑阴暗，风吹树叶沙沙作响，林中深处不时传来一阵低沉的鸣叫声，那是草鸮发出的声响，它们用泛着蓝光的眼睛仔细扫描着林间的一切。草鸮特别喜欢的目标是夜间出来觅食的小型啮齿类动物，食谱中还包括青蛙、蛇和大一点的昆虫。鸮是夜行猛禽中比较凶猛的一类，喙尖而勾，爪大而锐。它的叫声注满阴森和凄凉，仅以扑食时发出的尖厉叫声，就会把被捕者吓个半死！

Hiding under the night

The night in the forest is gloomy and horrible. Weak animals will choose to hide. For example, steep rock walls, tree tops and holes, caves, and dense bushes are all perfect shelters. Because the night in the forest is more dark and gloomy, the wind rustles the leaves, and there are strange noises in the silence. In the distance, there is a series of low calls. It was the sound of a grass owl, and the blue eyes were scanning everything in the forest. Their favorite targets are small rodents that come out at night to forage. Frogs, snakes, and larger insects are also included in their diet. All kinds of owls are the more ferocious species of nocturnal raptors, with a pointed and hooked beak and large and sharp claws. Its screams are as gloomy and desolate as a ghost, and the shrill screams when they prey on animals will scare the arrested to death!

高原上的清道夫

在地球生物圈的范围内,都有生命的存在,它是地球上生命的摇篮,也是地球生物的运动场。各种猛禽的巡视目的不同,金雕主要猎食草兔、鼠兔、旱獭、幼岩羊等活体动物,而兀鹫负责清理残渣,雪豹、金雕等吃剩的动物尸骨就是兀鹫的美食,从而间接起到了清洁环境的作用。如果是病死动物的尸体,只要是被兀鹫等食腐猛禽清理过,也能起到消除传染源的作用。兀鹫大餐之后,接下来就轮到一群群乌鸦来分享残渣剩饭了。"嘎……嘎……嘎……"一些乌鸦边鸣叫,边进食,地上的残渣很快便被它们打扫得干干净净。再小一点的东西,就该让蚂蚁来仔细处理了,物尽其用啊!大自然就是这样合理安排的,它们为了保洁、环卫而存在,为了生命延续而坚守。

Highland scavengers

Life exists everywhere within the"biosphere". It is the cradle of life on earth and a playground for earth creatures. Different prey birds have different patrol purposes. Golden eagles mainly prey live animals such as grass rabbits, pikas, marmots, and baby bharal. While the vultures are here to clean up the debris. Leftover animal bones of snow leopards and golden eagles are the food of vultures, which indirectly cleans the environment. Cleaning up the dead bodies with illness might help eliminate the source of infection. . After the vulture's feast, a group of crows come to share the leftovers. Some crows quack and eat while crying, and the debris on the ground is quickly swept away by the crows. Even smaller leftover things will be handled carefully by ants, thus make the best use of them! Nature is arranged in this its order. They exist for cleaning and sanitation, and they persist for the continuation of life.

兽类的大舞台
The Big Stage of Beasts

鼠类将伴随人们走向远方

  曾经被老鼠骚扰人们，都希望居住在没有老鼠的地方。然而，数量和人类相当的老鼠消失后，对人类有好处吗？鼠类是遍布全球的动物，种类多，适应性强，繁殖率高，是哺乳动物中最大的群体。

  从疾病的传播途径来看，如果地球上没了老鼠，细菌和病毒就会寻找其他宿主。病毒是适应性和变异性很强的病原体，当它失去一种传播途径后就会产生变异，以便找到新的宿主与感染途径。比如禽流感以前只在鸟类中传播，现在变异成可传染人的甲型流感，该病毒就是典型的宿主转换病毒。

  从一定意义上讲，鼠类的存在，对保持自然生态系统的稳定，也能起到一些积极的作用，比如生活在高原的鼢鼠和高原鼠兔，就维系着高原草甸的生态系统。

  生活在地球上的生物，无论以什么形式出现，其实都有各自存在的生物学意义。人类应该改变自己的思维习惯，不能凡事都以自身的视角去看待周围的事物，随便定义何为益兽，何为害虫，等等。鼠类虽然带有一些病菌，但它们却在实验室里为人类的健康付出生命！

  老鼠处于食物链的中间环节，它们在互相联系、互相依赖、互相制约的生态系统中扮演着重要的角色。在下，它们吃昆虫和植物、转移植物种子；在上，又是掠食性动物的食物。缺失了老鼠，大自然的生态系统则会陷入紊乱，失去平衡。

Rodents will accompany people forever

People who have been harassed by rats all want to live in a place where there are no rats. However, will it be good for humans after the disappearance of rats with the same number as humans? Rodents are animals all over the world, with many species, strong adaptability and high reproductive efficiency. They are the largest group of mammals.

From the perspective of disease transmission, if there are no mice on the earth, bacteria and viruses will find other hosts. A virus is a pathogen with strong adaptability and variability. When it loses a way of transmission, it will mutate and find a new host and infection route. For example, avian influenza used to only spread among birds, but now it mutates into Influenza A that infects humans. This virus is a typical host-transformed virus.

The zokor and the plateau pika living on the plateau maintain the ecosystem of the plateau meadow.

Creatures living on the earth, no matter what form they appear, actually have biological significance. Humans should change their thinking habits, look at the surrounding things from their own perspective, and define them as beneficial animals or pests. Rodents carry some germs, but they pay their lives for human health in the laboratory!

Rats are in the middle of the food chain. They are an important part of an ecosystem that is interconnected, interdependent, and restrictive. It eats insects and plants, transfers plant seeds, and feeds predators. Without the rat, the ecosystem will fall into disorder and lose its balance.

The big tail elf gnawing pine cones

Due to the different species, there are tree squirrel, rock squirrel, and ground squirrel. People have the impression of a squirrel with its big furry tail wagging around, holding pine cones and chewing on a tree. Rock squirrels are more widely distributed in broad-leaved forests and mixed coniferous and broad-leaved forests in North China and other places. Red-bellied squirrels live in some regional parks and surrounding areas of cities, and they are also the most common wild animals to be seen by urban residents in Southern China. Squirrels are active throughout the year, do not hibernate, and spend 70% of their time in foraging activities, foraging and storing food in coniferous forests. In addition to eating nuts, squirrels also eat some leaves, small insects, etc. They are used to storing nuts scattered on the ground and fungi on branches. They also need to turn the stored things in the sun when the weather is good.

啃松果的大尾巴精灵

由于种类的不同，松鼠被细分为树松鼠、岩松鼠、地松鼠。人们印象中的松鼠，普遍是长着一条摇来摇去、毛茸茸的大尾巴，用前肢抱着松果啃食，不断在树上跳跃的情景。在我国华北等地的阔叶林和针阔混交林中，分布最为广泛的是岩松鼠。赤腹松鼠多出现在一些地区的公园及城市周边地带，也是我国南方城市居民最易见到的野生动物。松鼠全年活动，不冬眠，70%的时间用于觅食，最乐于在针叶林中觅食和贮藏食物。松鼠除了喜食坚果外，也进食一些树叶、小昆虫等，习惯把坚果分散贮藏于地面，将真菌贮藏于树枝上，天气好时，还乐于把储存的口粮拿出来翻晒。

隐蔽在林间的"收割机"

在广袤的森林里，聚集着小麂、林麝、毛冠鹿、苏门羚、草兔等食草动物，它们采食着丰茂的植物，如果没有人类的打扰，它们将过着"丰衣足食"的安定生活。

在黄昏和黎明时分，小型食草动物会到溪边饮水，此时最容易观察到它们的身影。因为是小型食草动物，没有什么防卫能力，所以它们都生性胆小，警觉敏锐，养成了逃跑和躲避的习惯。

小型食草动物是食物链中的一个重要环节，也是能量流动的一部分。它们在森林里"收割"，也在森林中成长。

The "harvester" hidden in the forest

In the vast forest, Chinese muntjac, forest musk deer, tufted deer, serow, grass rabbit and other herbivores gather to feed on lush plants. If there is no human interruption, they lead a stable life of "ample food and clothing".

It's quite easy to observe the herbivores drinking water behind the stream at dusk and dawn. Because they are small herbivores and have no defense capabilities, they are naturally timid, alert and sensitive, and have developed the habit of escape and avoidance.

Small herbivores are an important part in the food chain and energy flow. They "harvest" in the forest and grow up in the forest.

高山里的"盾构机"

高原的清晨,地毯般的草甸一望无垠,嵩草和莎草是这里的优势种,还有一些美丽的龙胆与狼毒。远处传来鸟儿般的叫声,那是旱獭发出的鸣叫。呆萌的喜马拉雅旱獭,三五成群站立在洞口边,整齐地面朝东方冉冉升起的太阳,前肢合十,好像在感恩苍天又赐予了阳光明媚的一天。

旱獭喜欢打洞,它们在地下不知疲倦地挖掘,修筑着长长的地下长廊。在地下的洞穴中,旱獭修建了功能齐全的"宫殿",有储粮间、寝

室、活动房、卫生间、多处逃逸出口等。洞穴是旱獭赖以生存的空间，打洞是旱獭的重要工作。只要生命还在，"盾构机"就不会停歇！

在草甸上，任何一处空间都不是弱者生存的地方。天空翱翔的金雕、鵟和傍晚出没的鸮，随时都会把老弱病残者彻底消灭。要生存就需要强壮和智慧！

"Shield machine" in the mountains

In the early morning on the plateau, the carpet-like meadows are endless. Kobresia and sedge are the dominant species here, as well as some beautiful gentian and radix euphorbiae lantu. There was a bird-like call in the distance, which was the call of a marmot. The cute Himalayan marmots stand in groups of three or five near the entrance of the cave, facing the rising sun in the east, with their forelimbs folded together, as if they are grateful to the sky for giving a sunny day.

Marmots like to burrow. They dig tirelessly in the ground to build a long underground corridor. In the underground caves, the marmot built a fully functional "palace" with grain storage rooms, dormitories, mobile rooms, toilets, and multiple escape exits. Caves are the space on which marmots live, and burrowing is an important "work" of marmots. As long as life is still there, the "shield machine" will not stop!

There is no place for the weak to live on the meadow. The golden eagles and buzzards flying in the sky and the owls that haunt in the evening will wipe out the old, weak, sick and disabled at any time. To survive requires strength and wisdom!

与大牌同名的小熊猫混成了九节狼

小熊猫命名早于大熊猫,当人们发现大熊猫后,为了方便辨别,小熊猫只能让出美名。因此,在动物界也要离旗舰物种远一点,不然就会吃大亏。

小熊猫主要分布在我国西藏、云南、四川等地区的针阔混交林或常绿阔叶林中。它们善于攀爬,往往能爬到高而细的树枝上休息或躲避敌害,有时还将两脚下垂高卧于树枝上。小熊猫生性胆小、温顺,却是颜值很高的动物,非常讨人喜欢。

小熊猫和大熊猫虽一字之差,却是两种截然不同的动物。小熊猫是大熊猫同一生境的伙伴,竹子是共同喜好的美食。然而,大熊猫的名气太大,小熊猫的关注程度,被大熊猫的光环大大遮蔽。

The red panda with the same "surname" with giant panda

Red pandas are named earlier than giant pandas. When people find giant pandas, in order to make it easier to distinguish them, they can only give their names to them. Therefore, in the animal kingdom, stay away from the flagship species, or you will suffer a big loss.

Red pandas are mainly distributed in coniferous and broad-leaved mixed forests or evergreen broad-leaved forests in Tibet, Yunnan, and Sichuan. It is good at climbing, and can often climb on tall and thin branches to rest or avoid enemies. Sometimes the feet hang high on the branches. Red pandas are timid and

"森林萌主"
身边的
那些事
STEP INTO
GIANT PANDA
HABITAT

docile. Red pandas are also very attractive animals.

Although there is only one word difference between the red panda and the giant panda, they are two completely different animals. The red panda is a companion of the giant panda in the same habitat, and bamboo is a common delicacy. However, the popularity of giant pandas is growing day by day, and the attention of red pandas is greatly obscured by the halo of giant pandas.

树洞的实用价值远大于"虫洞"的未来意义

斑驳的阳光透过冷杉和云杉的枝梢洒向山谷，处处弥漫着祥和的暖意，很多动植物就生活在这样的氛围中。生长几百年的树木枯萎后，大大小小的树洞，就成了动物们相互争夺的稀缺资源。因为树洞既可以成为粮仓，也可以作为居所。在动物世界里，把树洞当成生活的港湾，可以成为挡风避雨、隐蔽藏身、生儿育女的摇篮。树洞为大熊猫、黑熊、鼯鼠、猫头鹰、松鼠提供了理想的住处，还为动物们的繁育和休息提供了最佳场所。特别有意思的是，一只松鼠和一只鸮会为一个树洞展开激烈的争夺战，往往互有胜负。

The practical value of the tree hole is far greater than the future significance of the "worm hole"

The dappled sunlight sprinkled on the valley through the branches of fir and spruce, filled with warmth everywhere, and many plants and animals lived in the forest. After hundreds of years of trees withered, big and small tree holes have become scarce resources that animals compete for, because the tree hole can not only be used as a granary, but also as a shelter for animals. In the animal world, the tree hole is regarded as a harbor for life, playing a role in sheltering from wind and rain, hiding, and cradling for childbirth. It provides shelter for giant pandas, black bears, flying squirrels, owls, and squirrels. The tree hole is the best place for animals to breed and rest. What is interesting is that a squirrel and an owl will fight fiercely for the tree hole, cannot expect the winner.

森林里的"推土机"

野猪是一种环境适应性极强的中型哺乳动物,地区不同,体态大小也有所不同。野猪食性极杂,几乎没有不吃的东西,一般来说,植物占食物的90%,多以嫩叶、坚果、浆果、草叶和草根为主。野猪凭借灵敏的嗅觉,能嗅到隐蔽的鸟巢,所以,它们有时会偷食鸟卵,特别是松鸡、雉鸡的卵和雏鸟。此外,它们还捕食兔子、老鼠,甚至包括蛇。

野猪身强力壮,头部较大,身体后部稍小。眼小,视力差,长而直的鼻子具有敏锐的嗅觉。犬齿发达,獠牙突出。

坚硬的鼻子是野猪的利器,它能像推土机一样拱翻土壤,从地下挖掘植物的根和球茎。它们在掘地觅食的时候,会把落叶、泥土和空气搅动在一起,其觅食行为,加速了森林凋落物的分解过程,改变了土壤组成、提升了肥力。而最重要的作用是重塑了林下环境,使掉落在地面的各种籽实有了萌生的机会。此外,被它拱翻的局部土壤又会给林下其他植物提供新的机遇,也给昆虫等小动物营造了良好生境。冬天,野猪大面积拱地觅食,清除厚厚的积雪,间接为鸟类和小型哺乳动物提供了觅食机会。

"Bulldozers" in the forest

The wild boar is a medium-sized mammal with strong environmental adaptability. The sizes produced in different regions are also different. Its eating habits are mixed, and there is almost nothing that cannot be eaten. Generally speaking, plants account for 90% of food and feed on young leaves, nuts, berries, grass blades and roots. Sometimes they eat bird eggs, especially grouse, pheasant eggs and chicks. The wild boar can smell the hidden bird's nest by its sensitive sense of smell. It also hunts rabbits, mice, and snakes.

The wild boar has a strong body with a larger head and a slightly smaller back part. He has small eyes, poor eyesight, and a long, straight nose with a keen sense of smell. The canine teeth are well developed and the fangs are prominent.

The hard nose is its sharp weapon. It can overturn the soil like a bulldozer and can dig roots and bulbs of plants from the ground. They agitate the fallen leaves, soil and air when they dig the ground for food. Foraging behavior accelerates the decomposition process of forest litter, changes the soil composition and improves fertility. The most important thing is to reshape the forest environment, so that all kinds of seeds that fall on the ground have a chance to germinate. The local soil overturned by it will provide new opportunities for other plants under the forest and create a good habitat for insects and other small animals. In winter, wild boars hunt for food in large areas and clear thick snow, which indirectly provides foraging opportunities for birds and small mammals.

高高山岗上的大牛

　　扭角羚也叫羚牛，是生活在高山上的大型食草动物，喜欢采食树叶、青草、种子和果实。多在早晨和下午觅食，中午休息。扭角羚在每年的3~5月份产子，有极强的护子行为，在觅食活动中也非常照顾幼子，经常喂乳、舔子。

　　扭角羚有集群现象，冬春季节聚集的数量多在60头以上，初夏开始以小组群活动。扭角羚有迁徙的习惯，冬季下移到低海拔地区，4~5月转移至河谷区域，6~7月又往高处移动。扭角羚有嗜盐的习性，常常采食带有咸味的矿物质。雄性扭角羚平时喜欢独居，所以被称为"独牛"。

　　扭角羚喜欢群居，整个团队由雌兽、幼兽和未成年兽组成。扭角羚在觅食时，会设置"哨牛"负责警戒，如发现异常情况，会即刻发出警告。迁移时，由幼子和亚成年兽走前，成年兽殿后。

　　扭角羚喜欢游戏，以头对头、角对角相撞，打斗游戏中会争取占领制高点，选择有利地形。扭角羚栖身于山势陡峭、树林茂密及多石崖和沟涧的地区，常常在高高的山梁上休息与活动。

"森林萌主"身边的那些事

STEP INTO GIANT PANDA HABITAT

The big cow on the high mountain

Takins are large herbivores that live in high mountains. They like to eat leaves, grass, seeds and fruits. They forage in the morning and afternoon, and rest at noon. Takins give birth in March to May and have strong child-protection behaviors. They also take good care of their young in foraging activities, such as breastfeeding and licking.

Takins are clustered. The number of takins in winter and spring is more than 60. In early summer, they start to move in small groups. Takins have the habit of migrating. They move down to low altitude areas in winter, move to river valley areas in April and May, and move to higher places from June to July. Takin has a salt-loving habit and often eats salty minerals. Males like to live alone, so they are called "solitary takin".

Takins like to live in groups and consist of females, cubs and juveniles. When takin foraging activities, "sentry takin" are responsible for alerting, and warnings will be issued if there are abnormalities. When moving, the adult takin walk after the young and sub-adult.

Takin likes games, head-to-head and corner-to-corner collisions. In fighting games, they all strive to occupy the commanding heights and choose favorable terrain. Takins live in steep mountains, dense forests, rocky cliffs and gullies, and often rest and move on high mountain ridges.

森林里的黄色旋风

如果你去过森林深处,你就会体验到风的巨涛,从远处到眼前,自头顶至远方,那摧古拉朽的气势定会让人刻记于脑海。在高山森林里,还有一种黄色旋风,那就是集群的金丝猴!

金丝猴是人类的近亲,共有川金丝猴、滇金丝猴、黔金丝猴、怒江金丝猴、越南金丝猴五个亚种,它们都以树芽、树叶、树皮、野果、鸟蛋等为食。

金丝猴营的群居生活,是以一雄多雌为组群,再由多个家庭构成金丝猴人群。其内部社会等级严密,金丝猴王国没有猴王,是由高等级家庭负责承担保护与领导猴群的任务。

金丝猴多栖于树上,个别时候下地觅食。在冬季,猴群会抱团取暖,负责警戒任务的金丝猴会爬上最高的树顶,哨兵必须具备舍自己为大家的社会责任感。川金丝猴迁移时,会上上下下不断跳跃、移动,宛如飘移的黄色旋风。

Yellow whirlwind in the forest

If you have been to the depths of the forest, you will experience the giant waves of the wind, from the distance to the front, from the top of the head to the distance, the destructive aura will be engraved in your mind. There is also a yellow whirlwind in the alpine forest, that is the cluster of golden monkeys!

Golden snub-nosed monkeys are close relatives of human beings. There are five subspecies, Sichuan snub-nosed monkey, Yunnan snub-nosed monkey, Guizhou snub-nosed monkey, Nujiang snub-nosed monkey and Vietnamese golden monkey. Golden monkeys all feed on tree buds, leaves, bark, wild fruits, and bird eggs.

The golden monkey lives in groups, with one male and multiple females as a family, and multiple families form a large group. The internal social hierarchy is strict. There is no monkey king in the golden monkey kingdom. High-level families undertake the task of protecting and leading the monkey group.

Golden snub-nosed monkeys live mostly on trees and feed on ground occasions. In winter, the monkey group will hug each other to keep warm, and the sentry monkey will climb to the top of the tallest tree which is responsible for alerting. The sentry must have a sense of social responsibility to give oneself to everyone. When the Sichuan golden monkeys migrate, they jump and move up and down like a yellow whirlwind.

称霸森林的"平头哥"二弟

与威震非洲草原的蜜獾"平头哥"相比,生活在中国大部分地区的鼬科动物们也不是等闲之辈!

鼬科动物是食肉目里最庞大的家族,各个生态位里几乎都有它们的身影。鼬科动物体型虽小,但十八般武艺样样精通的黄喉貂却敢挑战野猪,甚至还能威胁到大熊猫。

黄喉貂是陆生鼬科中唯一群居的种类,它们凭借灵活敏捷的身姿来去如风。黄喉貂的食物范围很广,从昆虫到鱼类、小型鸟兽,都在它的捕食目标之列,有时还要猎食猕猴、野猪、毛冠鹿、斑羚等。黄喉貂还有个与众不同的最爱——吃蜂蜜,故人们又称它为"蜜狗"。

另外,黄喉貂还有几个修炼"成精"的兄弟——偷鸡的黄鼬大仙、骑鸟背飞行的伶鼬和水中杀手水獭。

The brother of ratel who dominates the forest

Compared with the ratel in the African grasslands of Megatron, the weasels living in most parts of China are not idlers!

Weasels are the largest family in the Carnivora, and almost have their positions in every niche. Although small in size, the yellow-throated marten, who is proficient in 18 martial arts, dares to challenge wild boars, and even threatens giant pandas.

The yellow-throated marten is the only group living in the terrestrial weasel family. It has a wide range of food, from insects to fish, small birds and beasts, and sometimes also hunts macaques, wild boars, tufted deer, and goral. It has a favorite-like eating honey, so people also call it a honey dog.

In addition, it has a few "brilliant" brothers, the Siberian Weasel who "steals" chickens, the Least Weasel who rides a bird and the killer otter in the water.

善于群殴和掏肛的打手们

除了位居食物链顶层的猫科动物外,犬科动物更像是一帮打手。狼和豺都是犬科的常见动物,它们都喜欢群居,狩猎时当然也会采取"蜂群"战术。

在森林、在草地,猫科动物都是直接扑捉猎物,狼和豺均不能一招制敌,只能采取战术性围攻,通过不断追赶和骚扰猎物,待其烦躁、疲惫时,便使出极为"卑鄙下流"手段——掏肛!犬科动物一般猎取食草动物,其肛门是最薄弱的部位,狼和豺就选择从这里下手。

狼和豺虽然对老幼、病残、孤单的动物构成了严重威胁,但对食草动物种群的优胜劣汰、保持种群的旺盛活力却有一定的积极作用。

The thugs who are good at group fighting and anal digging

Except for cats at the top of the food chain, canines are more like thugs. Wolves and jackals are common animals in the canine family. They both like to live in groups. Of course, hunting also needs to adopt "bee colony" tactics.

In the forest and on the grass, cats directly capture their prey. Wolves and jackals cannot control the enemy with one move, but can only resort to tactical siege. Constantly chasing and harassing them, and when they are irritable and exhausted, they resort to "despicable and nasty" means to dig their anus! Canines generally hunt herbivores. Their anus is the weakest part. Wolves and jackals will start from here.

Wolves and jackals pose a threat to young, sick, and lonely animals. It has a certain positive effect on the survival of the fittest and maintaining the vigorous vitality of the herbivore population.

威武霸气的"雪山之王"

　　雪豹的生活区域基本是在高山雪线附近，在高海拔地区，也只有它是最凶猛的肉食动物了，所以被冠以"雪山之王"的称号。雪豹全身灰白色，布满黑斑，其尾巴几乎与身体等长。

　　雪豹是高原地区的岩栖性动物，经常在高山裸岩及寒漠带的环境中觅食岩羊、旱獭等。雪豹善于在陡峭的山崖间追逐猎物，长长的尾巴能起到了平衡、稳定、转向的作用，尾巴是雪豹的"神器"之一，它还会经常咬着自己的尾巴。

　　雪豹的存世数量极少，甚至不及大熊猫，目前已成为濒危物种。如果雪豹能在高原极寒的严酷环境里生存下去，那就意味着该区域食物链的完整性没有遭到破坏，脆弱的高原生态平衡就会持续下去。

　　在雪豹的周边，还有一个坚守雪原的兄弟，那就是兔狲。由于全球气候变暖，扰乱了各种动物的生境，不知雪豹和兔狲两兄弟还能坚持多久？

The mighty and domineering "King of Snow Mountain"

The living area of the snow leopard is basically near the high mountain snow line. In this high altitude area, it is the only ferocious carnivore, so it is called the "king of the snow mountain". The snow leopard is gray-white with black spots, and its tail is almost as long as the body.

Snow leopards are rock-dwelling animals in plateau areas. They often feed on bharal, marmots, etc. in the alpine bare rock and cold desert environment. Snow leopards chase their prey on steep cliffs, and their long tails play a role in balance, stability, and steering. The tail is the "artifact" of the snow leopard, and it is often found containing its own tail.

The population of snow leopards is small, even less than that of giant pandas, and has become an endangered species. In the harsh environment of the extreme cold plateau, snow leopards can survive, which indicates that the integrity of the food chain is not destroyed, and the fragile plateau ecological balance will continue.

There is also a brother who sticks to the snowy field around it, and that is manul. Global warming has disrupted the habitats of all kinds of animals. I wonder how long the snow leopard and manul can exist?

翱翔在天空的哺乳动物

鼯鼠是哺乳动物中的一种，头圆、眼大、吻短，耳壳发达圆宽，体型比松鼠略大，可依靠翼膜在森林里进行短距离飞翔。鼯鼠习惯在夜间的森林中活动，还会发出凄厉的叫声。鼯鼠身体两侧的翼膜状若鸟翅，远看很像一只飞翔的鸟，因此也称其为"寒号鸟"。

在动物界，蝙蝠是唯一能振翅飞翔的哺乳动物，与蝙蝠不同，鼯鼠只是靠翼形皮膜在空中滑行！

蝙蝠的视觉较差，而听觉则异常发达。在夜里，蝙蝠靠声波探路和捕食，当它发出的声波遇到其他物体时会自动返回，蝙蝠能据此判断出物体大小、移动还是静止、距离有多远。绝大多数蝙蝠是以昆虫和小节肢动物为食，其中的30%是以果实、花蜜和花粉为食，极个别种类以血液为食。

蝙蝠一般都有冬眠的习性，冬眠时新陈代谢的水平会降低。蝙蝠一般都居住在屋檐、树洞和幽暗的洞穴里，它们是一种与世隔绝的动物，会尽量避开与人类接触。

目前，一些洞穴探险活动和食用蝙蝠，严重干扰与伤害了长年隐居的蝙蝠。在蝙蝠生活的这些古老洞穴里，隐藏着许多秘密，它们就像潘多拉的魔法盒，人类已经触碰了人畜共患病毒的天然宿主！

Mammals soaring in the sky

Flying squirrels are a type of mammal, slightly larger than squirrels. Flying squirrels have a round head, large eyes, a short snout, and well-developed round ears. Flying squirrels rely on wing membranes to fly short distances in the forest. Flying squirrels move in the forest at night, they will make a sad-cry sound. The wing membranes on both sides of the body are like bird wings, and they look like a bird from a distance, so it is also called the "cold bird".

Bats are really mammals with strong flying ability. Bats are the only mammals that can flap their wings and fly. Unlike bats, flying squirrels only glide in the air on the wing-shaped membrane!

The vision of bats is poor, but the sense of hearing is abnormally developed. At night, bats use sound waves to find their way and hunt food. When the sound waves emitted by them encounter an object, the sound waves return. The bat can distinguish the size of the object, whether it is moving or still, and how far away it is. Most bats feed on insects and other small arthropods, 30% feed on fruits, nectar and pollen, and very few feed on blood.

Bats generally have the habit of hibernation, and their metabolism is reduced during hibernation. Bats generally live in eaves, tree caves and dark mountain caves. They are an isolated animal and try to avoid contact with humans.

At present, some cave exploration activities and eating bats have disturbed and harmed bats that have lived in seclusion for hundreds of years. There are many "secrets" hidden in these caves, like Pandora's magic box, humans have touched the natural host of zoonosis viruses!

第六章 Community of Life
生命共同体

Chapter Six

一般来讲，"生物圈"范围内生物都各有其生活区域，都各自遵守着生存的自然法则和规矩。但个别自然异常现象或人类干扰，往往会打破既有的稳定与平衡，并导致发生一些局部的、偶然事件。

Generally speaking, all creatures in the "biosphere" have their own living areas and abide by the laws and rules of nature. Sometimes, individual natural anomalies and human disturbance break the stability and balance, and cause some regional and accidental events.

人类应该努力了解自己生存的这个世界

地球不只有人类，世界是由众多的生命体组成。人类生活的环境，不仅有春光明媚、鲜花烂漫，还有风暴、洪水、地震、瘟疫等灾害。其实，人类始终与微生物相互共存，人体内与人体表面都生存着细菌、真菌和病毒。很多微生物其实是与人类和平共处的，一旦这种平衡被打破，受害的首先就是感知敏锐的人类，于是，大自然的所有生物包括人类便有了生生死死的生命轮回。

一切生命都应该感恩大自然，是太阳给了我们光照、温暖；是地球赐予了我们空气、水、土壤，在这适宜的环境条件下，才孕育出了彩色斑斓、多姿多彩的生物世界。

对人类而言，虽然我们自身的发展已经比地球上其他生物先进了很多，但在面对变幻莫测的自然环境时，我们仍然处于一个非常低微的地位，在大自然面前，无论人类多么强大，都显得非常渺小。事实告诉我们，虽然说时代的发展给我们带来了优质的生活，但我们仍然应该始终保持对自然的崇敬之心，要以科学的视角看待世界，这样才能了解这个世界，才能做到人与自然和谐共处。

Human beings should strive to understand the world in which they live

The earth has not only humans, and the world is composed of many life forms. The environment in which human beings live has not only bright and beautiful scenery, but also disasters such as storms, floods, earthquakes, plagues and other disasters. In fact, human always coexist with microorganisms. There are bacteria, fungi and viruses inside and on the surface of human body. Many micro-organisms coexist peacefully with human beings. However, once the balance is broken, human beings are the first victim. Therefore, when there is life, there is death.

All life should be grateful to nature. The sun gives us light and warmth, and the earth gives us air, water and soil. Only this suitable environmental condition can give birth to colorful and rich biological world.

Although development of human beings is much more advanced than other creatures on the earth, we are still in a passive position in the unpredictable natural environment. No matter how powerful human beings are, we are still very negligible in the face of nature. It tells us that although long time development has brought us a better quality of life, we should still respect nature, and strengthen the study of science and technology. We should view the world from a scientific perspective, so that we can understand the world and achieve harmonious coexistence between human beings and nature.

人类社会始终在与瘟疫的斗争中求生存

人类社会的发展与进化,也是适应自然、认知自然、应对自然的艰难历程。从原始人到现代人,地球上没有任何特殊类群会置身事外,这就是命运共同体!除了地质灾害、气候异变会威胁到人类的生存外,人类生存环境中的微生物,也每时每刻与人们长期共存,一旦这种平衡被打破,人类的生存同样会面临威胁。

地球的自然生态系统,本来为人类和动物提供了一套合理的"疾病限制机制"。自然界的生物多样性,通过"稀释"或缓冲效应,限制了许多病原体的扩张和影响。然而,最可怕的是,由于人类群体的不断扩张,一再违背自然规律,无限的物欲长期驱动着掠夺式的攫取,超出了自然界的承载能力。一旦这些作为打破了生物圈的平衡,灾害就会降临。在人类历史上,就多次遭受过瘟疫的侵害,曾造成上百万甚至千万人因染病而死亡,如伤寒、天花、麻疹、鼠疫、腮腺炎、麻疹、霍乱、疟疾、黄热病、西班牙型流行性感冒、俄国斑疹伤寒等。

地球不只是人类的地球,微生物也是若干生命体中的一份子,它们同样拥有谋求生存的权利。因此,人们要有充分的认识和思想准备,微生物会与人类长期共存。微生物灭绝之时,也是人类灭亡之日!

Human society has always been struggling for survival when fight against plague

The development and evolution of human society is a difficult process of adapting to nature, recognizing nature and coping with nature. From primitive man to modern man, there is no special group staying out of the process. This is the Community of Common Destiny! In addition to geological hazards and climate change that threatening human survival, microorganisms in human living environment also coexist with people all the time. When the balance between microorganisms and human is broken, human life will be threatened.

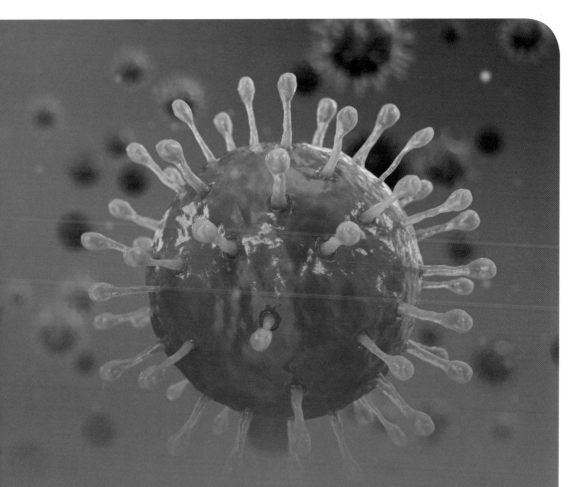

The natural ecosystem provides a "disease restriction mechanism". The biological diversity in nature limits the contact and influence of many pathogens through "dilution" or buffering effects. However, the continuous expansion of human groups, violation of laws of nature, unlimited material desire and long-term predatory utilization have exceeded the capacity of nature. At this point, the balance of the biosphere is broken, and disasters come. In the history, human beings have been attacked by pestilence, such as typhoid, smallpox, measles, plague, mumps, measles, cholera, malaria, yellow fever, Spanish influenza, Russian typhus, etc., for many times, which caused millions of people get sick and die.

The earth doesn't belong to human beings. Microbes are also part of the life entity, and they have rights to survive. Therefore, people should realize that microbes will coexist with human beings for a long time. When microorganisms extinct, the extinction of human is not far from us!

一样的冬季一样的寒冷

　　动物携带的细菌和病毒不仅伤害其自身,而且还屡屡感染人类。2002年冬季发生的(SARS)重症急性呼吸综合征,到2003年夏才被消灭,病毒源头是野生动物;2004年初发生的禽流感席卷美国和部分亚洲国家,上百万只家禽染病死亡,多人因感染禽流感病毒而去世;2005年、2006年又出现禽流感流行,病毒源头还是动物; 2019年初,非洲猪瘟迅速蔓延,不仅家猪发病,同时还传染到野猪; 2020年6月,阿联酋、内蒙古发现鼠疫患者,病毒源头同样是动物。

"森林萌主"身边的那些事
STEP INTO GIANT PANDA HABITAT

一样的冬季,一样的寒冷,动物和人类都面临病毒感染,共同处于一个危险的境地。其实,不同物种之间多数存在生物隔离,一般不会交叉感染,但近些年情况却有些不同,禽类、食草和食肉哺乳动物所携带的疾病往往会传染给人类。这到底是怎么了?非常值得人类认真思考!

The same winter, the same cold

Bacteria and viruses carried by animals not only attack themselves, but also infect human beings. SARS (Severe Acute Respiratory Syndrome) occurred in the winter of 2002 and was eliminated till mid-2003. The source of the virus was wild animals. At the beginning of 2004, bird flu swept across the United States and some Asian countries. Millions of poultry died of the disease. In 2005 and 2006, bird flu epidemic occurred again, and the source of the virus was still animals. At the beginning of 2019, African swine fever spread rapidly to not only domestic pigs, but also wild boar. In June 2020, plague patients were found in UAE and Inner Mongolia, and the virus was originated from animals.

The same winter with the same cold. Animals and human beings are exposed to virus infection, and are in a dangerous situation together. In fact, most species are biologically isolated and normally do not cross-infect. However, in recent years, diseases carried by poultry, herbivorous and carnivorous mammals have often been transmitted to humans. What happened? It is a thought-provoking question!

人类还能称霸地球多久？

地球是一个物种丰富的生命世界，有着38亿年的生命史，在漫长的岁月里，诞生了多种多样的生命。

在地球漫长的地质年代里，时而会出现霸主级的生物，如3亿多年前的巨虫时代，称霸地球的是现在非常渺小的昆虫。那个时期的蜻蜓，翼展达1米多，蜈蚣的体长达到3米。那时的昆虫力气很大，还有厚厚的盔甲，在竞争中常立于不败之地，昆虫凭此称霸地球。

昆虫时代之后，才是恐龙称霸。恐龙体长可达数十米，高十多米，依靠庞大的体型，恐龙自然成了地球的新霸主，并统治地球长达1.6亿年，只因一颗小行星偶然撞击地球，才终结了恐龙的霸主时代。

整个地球上的生命又经过了数千万年的演化，才出现了原始人类。人类不断进步，最终依靠自己的智慧树立了地球新霸主的地位。

未来有三种情况可能会终止人类的统治：一是地球上有很多生物在继续进化，智慧动物与微生物是潜在的威胁；二是人类自己，地球人习惯把其他生命体都看成是"薯条"，贪婪的欲望暴露了人的本性，无限制的掠夺将把人类自身送进万劫不复的境地；三是外星生物的入侵和宇宙中流星的撞击，都有可能随时终结人类的霸主地位！

How long can human dominate the earth?

The earth is a living world that is rich in species, and has a 3.8 billion years history. Various kinds of lives were born in this long time.

In the long geological years of the earth, sometimes there are hegemonic creatures. For example, in the Carboniferous period which is more than 300 million years ago, the earth was dominated by very few insects. At that time, dragonflies had a wingspan of more than one meter and centipedes were up to three meters long. Insects had great strength and thick armor. They were always invincible in the competition, which made insects dominate the earth.

After the Carboniferous period, dinosaurs dominated. Dinosaurs can reach tens of meters in length and more than 10 meters high. Because of their huge size, dinosaurs naturally became the new overlord of the earth and ruled the earth for 160 million years. As asteroid accidentally hit the earth, the dinosaur era ended.

Life on the earth has evolved over tens of millions of years before primitive human beings appeared. With continuous progress, human beings have become the new overlord of the earth thanks for their wisdom.

There are three situations in the future that may end human domination. Firstly, if there are too many creatures on the earth evolving, intelligent animals and microorganisms will become potential threats. The second situation is about ourselves. People on the earth regard other living organism as "chips". Greed and unlimited plunder will put human beings into a state of eternal ruin. At last, the invasion of alien and the impact of meteors may end human supremacy at any time!

图书在版编目（CIP）数据

"森林萌主"身边的那些事 = STEP INTO GIANT PANDA HABITAT : 汉英对照 / 大熊猫国家公园都江堰管护总站编著. -- 成都 : 四川科学技术出版社, 2020.11
（"森林萌主"自然教育系列丛书）
ISBN 978-7-5364-9976-8

Ⅰ.①森… Ⅱ.①大… Ⅲ.①生态环境保护—青少年读物—汉、英 Ⅳ.①X171.4-49

中国版本图书馆CIP数据核字(2020)第218859号

大熊猫国家公园都江堰管护总站　编著　（汉英对照）

出 品 人	程佳月
责任编辑	程蓉伟
出版发行	四川科学技术出版社
封面设计	李　庆
责任印制	欧晓春
设计制作	四川蓝色印象艺术设计有限公司
印　　刷	成都市金雅迪彩色印刷有限公司
成品尺寸	170mm × 240mm
印　张	12
字　数	100千
版　次	2020年11月第1版
印　次	2020年11月第1次印刷
书　号	ISBN 978-7-5364-9976-8
定　价	52.00元

■ 版权所有·侵权必究
本书若出现印装质量问题，联系电话：028-87733982